T0214254

SpringerBriefs in Applied Sciences and Technology

PoliMI SpringerBriefs

More information about this subseries at http://www.springer.com/series/11159
http://www.polimi.it

Francesco Augelli · Matteo Rigamonti ·
Paola Bertò · Alessandro Marcone

Preservation and Reuse Design for Fragile Territories' Settlements

The Anipemza Project

POLITECNICO
MILANO 1863

Francesco Augelli
DASTU
Politecnico di Milano
Milan, Italy

Paola Bertò
Milan, Italy

Matteo Rigamonti
Milan, Italy

Alessandro Marcone
Milan, Italy

ISSN 2191-530X   ISSN 2191-5318   (electronic)
SpringerBriefs in Applied Sciences and Technology
ISSN 2282-2577   ISSN 2282-2585   (electronic)
PoliMI SpringerBriefs
ISBN 978-3-030-45496-8   ISBN 978-3-030-45497-5   (eBook)
https://doi.org/10.1007/978-3-030-45497-5

This Springer imprint is published by the registered company Springer Nature Switzerland AG
The registered company address is: Gewerbestrasse 11, 6330 Cham, Switzerland

Contents

Chapter 1
Anipemza. The Birth of Interest on a Forgotten Extraordinary Fragile Site in Armenia

Francesco Augelli

Abstract Anipemza represent a unique study case in Armenia because of a singular settlement rich of history and an interesting architectural, environmental and landscape example. This book aims to describe, step by step, the reconstruction of the historical period when the village was settled and the life of the village, its values' definition and the consequent urgent necessity of preservation intervention and revitalization.

Keywords Armenia · Anipemza village · Yereruyk Basilica · Architectural preservation · Adaptive reuse · Fragile territory

This book born after Specialization School on Architecture and Landscape Heritage (SSBAP) thesis at Politecnico di Milano, discussed in December 2017 by Paola Bertò, Alessandro Marcone, Matteo Rigamonti and Angelo Rossi.

The idea of thesis, entitled The Anipemza Project, was decided between September and October 2014, during a SSBAP Armenia study tour: while visiting Yereruyk basilica archaeological site, 102 km far from Yerevan, I introduced some students to Anipemza village as a possibility of an interesting thesis topic. I visited first time Anipemza on 27th April 2013 during a survey with two Armenian students, Artin Muradian and Laura Yeghiyan, of Postgraduate 1st level University Master in "Architectural restoration site supervisor in Armenia" held in the year 2012/2013 and where I was Master director and professor of Architectural Preservation and Reuse design from 2011 to 2014 [1].

The 2013 visit to Yererouyk Basilica's Archaeological site was organised with the purposes to check the problems and logistic because my students were starting to work there for their thesis. So, I asked to visit the near (only 650 m from the archaeological site) village visible from the Basilica area. The first impression, while entering in the village, was amazing. It appeared as a forgotten, asleep over time village, not born spontaneously as many others vernacular villages visited in Armenia, but with a well-studied and organized urbanistic and with a very high quality of architecture. Students were surprised about my astonishment, thinking they I was joking because saying continuously "amazing site…". I asked to my students to wonder to some citizens which kind of village it was, in the middle of a sort of desert. A gentle lady

Fig. 1.1 View from the road of the panorama: on the left Anipemza, in the middle the Yereuyk Basilica area, in the background the tuff stone mine

explained us an impressive and rich history, telling us about children hosted after Armenian genocide, political prisoners, quarry workers and much more.

In the same year I have been there again with my first armenian student in Politecnico di Milano, who later became the Director of Monuments of Armenia, Armen Abroyan, to show him that amazing but decayed village with the aim of save it and to make it known in Armenia and worldwide. As it be normal for a very kind people as Armenians are, a family organized, during this second survey, a fast lunch offering us everything of better the family had. After that two visits others have been organized between 2013 and 2014 with other two my Master students: Arin Khachatourian Saredehi and Lousineh Khachatourian Saredehi. Their support was indispensable for me, also because no one of the citizens spoke English at all. So, they not only helped me to search and translate historical documents but also to do many interviews to the eldest citizens of the village. The Mayor and the citizens, surprised of our interest and passion about their village, were really kind and available and they accepted not only to be interviewed by me about past life in the village and about the actual problems, but also allowed me to access and copy many historical photos from their family photobooks. The result of this first step of research and work have been the first research published in 2015 on the Scientific Papers of National University of Architecture and Construction of Armenia signed by me and by my students Arin and Lousineh [2] (Fig. 1.1).

But this was only the first step of the work with the final aim of do proposals for preservation, reuse and valorisation of the village together with Yereruyk Basilica archaeological site. Pursuing this desirable aim required many efforts, and then the idea of a thesis, explained in short in this book, was born.

In early Spring 2015 my Specialization School Students, helped by me, Arin and Lousineh, started with the surveys and data collection useful for the thesis. From the first steps moved into this small but strongly characterized town we understood that there were interesting architectural qualities in these local stone buildings, difficult to discern from afar, but neat and evident once in.

The second, more important, impression was, as said before, to have been pushed back in time in an extraordinary place: a company town of the first decades of the Twentieth Century where the soviet urbanism was clearly narrating enthralling stories of a different lifestyle concept than the one our western European society actually used to live. The fact that the analysis the authors were undertaking was moving from the mere building preservation's boundaries and flowing into more society-related studies definitely convinced me that a thesis on this topic would have been a fascinating and somehow responsible challenge to take. It was just few months later, coming back to the village for surveys, when the authors actually realized the conditions of the people still living the village from the hygienic and occupational point of view and of their harsh living standards. The impression of a nearly abandoned village was now confirmed, such as the urge of a reuse project that could cope at the same time with a progressive preservation of the town buildings, but, more importantly, suggesting a masterplan reactivating a local virtuous microeconomic system through interconnected reuse strategies working on the existing touristic travel routes [2].

Knowing very well the many limits of this project proposal, the authors sincerely hope that it will raise attention to this somehow limited in size but still important social emergency, possibly suggesting some valid food for thoughts for any further intervention in aid to this village. If Anipemza will be forgotten the same will happen to all the many different social and historical values intertwined to its memory, whose complexity this work will try to exploit. Most of the historical information about Anipemza come from the interviews of the elder people and this book is the first one who tell a so rich and interesting history avoiding it will be lost forever. Unfortunately, many of the interviewed elder passed away in the last years and with them the witnessing of an amazing and dramatic history. The warmth of the touching welcoming spirit of the simple but at the same time dignified Anipemza citizens and authorities alike, demonstrated during the in-town visits, is something that will be preserved in the authors' hearts way beyond the thesis and book-writing times (Fig. 1.2).

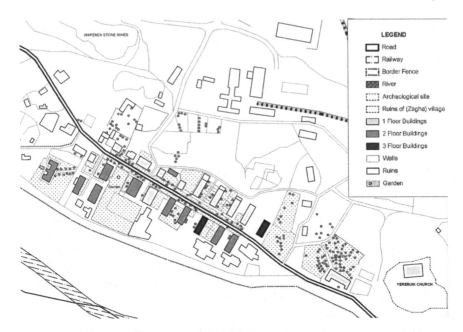

Fig. 1.2 Analysis of Anipemza's buildings. This was the one of the first maps realized after the first surveys performed with Lousineh and Arin Khachatourian Saradehi in 2014

Acknowledgements We wish to thank, for their kindness and availability, the Mayor of Anipemza Dr. Harutyun Tarlanyan, the Municipality leading specialist accountant Mr. Arsen Davdyan, all the wonderful people of Anipemza. Also, many thanks to Ing. Artin Muradian and Ing. Laura Yeghiyan and specially Arch. Arin Khachatourian Saradehi and her sister Arch. Lousineh Khachatourian Saradehi and Marco Germi.

Special thanks to Angelo Rossi who decided not to participate in the publication of this book due to lack of time but that for us authors is ideally co-author of the text.

References

1. Augelli F (2014) An international multidisciplinary cross-cultural cooperation project-the master's organized near YSUAC: sharing culture and knowledge. In: Casnati G (ed) The Politecnico di Milano in Armenia. An Italian ministry of foreign affairs project for restoration training and support to local institutions for the preservation and conservation of Armenian Heritage, pp 77–89. OEMME EDIZIONI, Venezia
2. Augelli F, Khachatourian Saradehi A, Khachatourian Saradehi L (2015) Anipemza: from genocide orphans' village to workers village. First proposals for conservation, valorisation and improvement of an interesting architectural settlement example and of a rich history site in Armenia. Scientific Papers of NUACA IV(59):14–28

Chapter 2
Historical Analysis and Anipemza Village's current condition

Francesco Augelli and Alessandro Marcone

Abstract Anipemza, in Shirak province, is located on the border of Armenia with Turkey, on the bank of Akhurian River on which opposite side there are, not so far, the ruins of the ancient and famous Armenian city of Ani, in Turkey since 1915. The old Anipemza, founded in IV century by Kamsarakan princes, is a village and rural community (municipality) of Armenia. Anipemza's territory is mainly well known for the near ruins of Yererouyk basilica of four–five century and for its quarry. The "new" village is characterized by an urban layout of particular interest and the impression is that to be in a workers' village of the early twentieth century. Near the buildings is possible to see a wide-open quarry for the extraction of the tuff and pumice. We cannot exclude that with the tuff of this quarry was built the Yererouyk basilica. The oral witness obtained by interviewing the residents have established that Anipemza, since 1926, was a village for the orphans of 1915 genocide and then has been also a penal colony for dissidents' forced labour during the Soviet regime in Armenia. The village is characterized by two distinct areas separated by a tree-lined entrance oriented towards the south north. The area problems are mainly: the lack of work for the inhabitants, repair of the water supplying system (repair of drinking water), lack of bathrooms, preservation and enhancing of the buildings, repair of road paving, agricultural products sales and completion of natural gas infrastructures. However, the very high urban and architectural quality of Anipemza deserves a contribution in order to make it known expecting that it will be preserved and valorised.

Keywords Armenia · Shirak region · Anipemza village · Yereruik basilica · Armenian's genocide · Orphans · Sovietization · Company town · Quarry · Tuff stone · Fragile territory · Abandonment

2.1 Area Introduction

Anipemza is a village located along the Turkish boundary in the North-Western part of Armenia, in the Shirak region of this small, vivid, southern-Caucasian country characterized by a plurality of deeply various and charming natural landscapes set on the Armenian plateau. The border with Turkey is represented by the gorge created

© The Author(s), under exclusive license to Springer Nature Switzerland AG 2021
F. Augelli et al., *Preservation and Reuse Design for Fragile Territories'*
Settlements, SpringerBriefs in Applied Sciences and Technology,
https://doi.org/10.1007/978-3-030-45497-5_2

by the Akhurian river, springing from northern lake Arpi and flowing into the southern Aras river. The village is set few hundred meters on the Eastern side of this natural rift.

The landscape here is characterized by a honey-blonde rocky plain with rather low and sparse vegetation, surrounded North and North–East by the beautiful spurs of the Aragats range. Even if the surrounding landscape appears to be beautifully solitary and natural, suggesting isolation at first sight, the village is set less than a couple of kilometres from one of the principal Armenian highways and railways, connecting the capital Yerevan with Gyumri, the second Armenian city in size and influence (Fig. 2.1).

A few hundred meters from the village the monumental ruins of the Yereruyk Basilica, which is one of the best examples of architecture of fifth century and one of the most important monuments of Armenia (Fig. 2.3). The cathedral was made of red tuff, materials extracted from the nearby quarry have been used, but in nowadays it is semi ruined. It is a part of a big complex, but the relation between the cathedral and other elements of the complex is still a mystery. Although archaeologists have done surveys for about one century on the complex, the results have not been completely published yet [1] (Fig. 2.2).

Fig. 2.1 Political Map of Armenia (from: Nations Online Project)

Fig. 2.2 Views of Anipemza village surroundings. General view from south east (top) general view to east (middle). View of the entrance of the village from south-east (bottom)

Fig. 2.3 Ruins of the Yereruyk Basilica and archaeological site

Other structures around the building are dedicated to Bagraduni[1] [2] period (tenth century). There is also an Iron Age necropolis nearby with monuments and some jewellery related to 2nd millennium BC was found.

It is also very important to mention that the Ani[2] [2] archaeological complex, which for Armenia represent the ancient capital ruins, is located just a few kilometres North of the village, on the opposite side of the Akhurian gorge, in Turkish territory.

2.2   Location and General Data

Anipemza is situated 49 km far from the centre of Shirak province, at 185 km from Yerevan to southeast and 50 km from Gyumri town to north.

Anipemza municipality occupies an area of 6, 7 km^2.

It is only 300 m far from the state border of Armenia with Turkey and about 15 km crow flies from the ancient site of Ani, the old Armenian capital, situated on the left side of Akhurian River, on a flat area on 1430 m on the sea level and which has been included in Turkish territory since 1920.

The only point of view, from Armenia, to Ani is in Kharkov, a village far from Anipemza a few kilometres, but it is impossible access there without permission. Anipemza was included in the province of Yerevan, district of Alexandropol (then Gyumri). In 1938 Anipemza became town-borough in Shirak province.

After the reform of 1995 it is known as village as rural series.

The climate in the village is temperate mountainous.

The winter is long and cold with constant snow cover.

The summer is hot and moist. Annual precipitation amount is 500 to 600 mm.

The highest temperature in summer is 30 °C and the lowest temperature in winter is −20 °C. Downfalls amount is 450 mm.

Natural landscape is black soil valley. Climate and geographical situation are like a foothill. Migrants from nearby villages founded the "new" village in 1926.

---

[1]The Bagratuni family of Armenia will become known since the first century BC when they still served under the Artasside dynasty. Ashot I was the first Bagratide King, the forefather of the royal dynasty, as it was recognized as such by the court in Baghdad in 861 A.D., which provoked the war with the local Arab emirs. Ashot won the war and consequently was recognized King of the Armenians also by Costantinople in 886 B.C. Ashot III, named "the Compassionate", transferred the capital city of Armenia Kingdom from Kars to Ani, which fell into the hands of Byzantines in 1045 B.C. This dynastic line end in 1118 A.D. with the occupation of Lori by the Muslim troops of the Selgiuchids. The members of the Bagratuni family survived, however, until a later date, though losing any political importance.

[2]Ani is a medieval city, today in ruins, which is placed in the Turkish province of Kars, near the boundary with Armenia During High Middle Ages it was the capital city of the Armenian kingdom, which included the vast majority of present Armenia and the eastern Turkish. At the time of its greater development it was known across all the region due to its splendour and prosperity, which allowed Ani to compete with cities like Constantinople, Cairo and Baghdad; After the conquest by the Mongolian army (XIII sec.), and later the Ottoman (XVI sec.) it was progressively abandoned and forgotten for centuries.

Some of the ancestors of the inhabitants have moved here in different periods from Armenia: Talin, Ani, Artik, etc.

In 1831 there were 647 residents, in 1873 they were 2359, in 1931 they were 529, in 1959 they were 1119, and in 1979 they were 558.

The national statistics community of 2013 said that on January 1st there were 392 people in municipality territory of Anipemza.

Men represent the 49% and women the 51% of the whole village's population.

The age percentage of the population is: children 32%, able to work 53%, adult 15%. The occupied area is 340 ha.

The ones for pastures are 220 ha.

There are pumice and tuff mines.

Population is nominally 520 people (live there 392), 120 families.

The population occupation is ranching and seasonally mines.

Near to village there was an industrial complex of building materials.

Anipemza is rich of these materials and has been a specialized branch of industry in the economy advanced by building materials production.

They were supplying the products to different cities and Soviets countries too. Nowadays the activity of mines is limited to seasonal and few works.

Especially tuff stone, andesite, cement and pumice, of which it has the name "Pemza" near Ani.

Even the name Anipemza has particular significance for Armenians because it reminds their inaccessible ancient capital of Ani.

The available lands are as following: 10% is the living residential land area, while the industrial land is 12% (Figs. 2.4 and 2.5).

Fig. 2.4 View of the tuff quarry in the '60. (From private Anipemza's people photo books)

Fig. 2.5 Industrial complex near the village and mines of tuff stone in Anipemza in the '50. (From private Anipemza's people photo books)

The agricultural lands and large pastures are the 72% of the land (477 ha). Other occupations are animal husbandry (cattle and small livestock), beekeeping, and vegetable growing.

The village has a school, a library, a post office and a labourer's canteen but as it will be described later, main of this facility nowadays do not work (Fig. 2.6).

Fig. 2.6 Different views of Anipemza main road. The entrance to the village (top) coming from Yereruyk Basilica and the main road inside the village (bottom)

Fig. 2.7 Ruins of the Turkish village of Zagha in Anipemza

2.3 Anipemza in the Past

2.3.1 The World War I on the Transcaucasian Front

During the first years of the 1900s a great part of Armenia was geographically incorporated within the Ottoman Empire: still today in Anipemza it is possible to see the remains of vernacular architecture, referring to the presence of a pre-existing Turkish village named Zagha (Fig. 2.7).

In that time irreparable divisions came to be formed, as whole empires fell apart unleashing peoples and ethnic groups eager to acquire justice and independence.

Actually, it was a real political struggle for survival which involved, despite itself, Armenia: this one came to be contested between three different forces. The first force was formed by the imperial powers of middle Europe, which wanted to maintain their spheres of influence also in the Caucasian areas because of their strategic importance as the doors to Asia. The second one was the group of Bolshevik, men inspired by a Party, heirs to centuries of misery and humiliation, convinced that the revolution would change the social status bringing well-being and happiness. The third was the Turks who, in the name of the utopia of the Pan-Turkism[3] [3], decided to invest in the ultra-nationalism to avoid the fragmentation of their land. In this strategic game between different powers, the only ones who could not decide their own fate were precisely the Armenians, who were soon to be moved like pawns on a chessboard, involved in a game much bigger than them. So they suffered the war, the revolutions, and faint hopes (soon frustrated) up to the genocide. Because of the destruction of

---

[3]Pan-Turkism was born in the late nineteenth century in the empires of Central Europe and the Ottoman Empire, where the feelings of commonality of origin was strong among the Turkish peoples and the so called Turanian peoples. At the beginning of the twentieth century, there were political and economic factors that favored a rap-prochement (also cultural) between Austria-Hun-gary and the Ottoman Empire (it is the period of the railway line connecting Berlin-Vienna-Buda-pest to Constantinople-Baghdad and of the large attendance, by the young Ottoman aristocrats, of central European and Danube universities). The father of Pan-Turkism is thought to be the Hun-garian orientalist Ármin Vámbéry. The ideology of Pan-Turkism therefore seeks to promote the union of all Turkish peoples, including, at least in origin, the Hungarians. With the end of World War II and the Soviet occupation of Hungary, the Pan-Turkism was no longer in the agenda of the Hungaria

all documents by the Turks, for a long time was discussed if the Armenian genocide was carefully planned or if it was an improvised choice made in the early months of World War I, due to a succession of events [3].

It was 1908 when the Young Turks appeared in the scene of the Ottoman political life. The movement was essentially made up of intellectuals most of whom coming from Russia. They wanted to improve the economic and cultural levels of the Turkish society after centuries of Ottoman obscurantism, where poverty and illiteracy were the norm: this would have only been possible through the unification of all Turkish peoples under the name of their faith and their customs. In this first phase of the project, which included the Turkish reunification with the Tatar population of Azerbaijan, the main obstacle was represented by the Armenians, because after the sequence of the Ottoman army defeats during the Balkan war, the Turkish possessions and ethnic minorities in their territories were reduced to historic lows.

The Young Turk leaders were able to voice their reasons only since 1914 when, thanks to the Committee of Union and Progress, they had all the tools to execute a plan—structured, totalitarian and driven by a central commission—able to prepare and conduct, throughout the empire, the elimination of ethnic minorities.

On 8 February 1914 Russia and Turkey signed a treaty for the creation of the Armenian provinces in Anatolia. Anipemza was inserted in the administrative district between Trabzon, Sivas and Erzurum, while the other province contained the districts of Van, Bitlis, Diyar-bakir and Harput. Each of them had to have a foreign general inspector who was appointed to supervise the implementation of the reforms. This agreement seemed to mark the beginning of an alliance between Russia and Turkey; however, on 2 August Turkey signed a secret agreement with the German Ambassador Wangenheim for a military alliance with Germany in the event of eruption of hostilities [3].

At the outbreak of World War I, the Turks immediately poured 200,000 regular soldiers into the Russian front: these men found themselves in an area inhabited by Armenians and this greatly increased the hatred for the latter.

In Erzurum it was immediately convened a conference where the Turkish Unionist Party asked the Armenian Revolutionary Federation (ARF) to provoke an uprising of the Armenians of Russia to facilitate the penetration of the Ottoman troops in Transcaucasia: as a reward, an autonomous state which would include the Armenian territories geographically inside the Russian borders would be granted.

The three Armenian leaders rejected the proposal, stressing the neutrality of their party, but ensuring maximum loyalty if the conflict would reach the Turkish territory; actually, once mobilized the Ottoman army, many Armenian citizens joined the army and very rare were the cases of rebellion or desertion, much fewer than those of the Turks [3]. Nevertheless the same will happen within the Russian army, where many Armenians (Russian citizens) took part in the Zar's army as volunteers. At the outbreak of the conflict, Russian forces (including 4000 Armenians) crossed the borders and broke through the enemy lines up to Sarykamysh, where they met a violent Turkish resistance. There, on 22 December a Turkish counter-offensive which defeated the Russian troops in a few days took place. The Turkish commander, however, had not planned to make provisions for the harsh and cold winter of the

Armenian highlands: with inadequate clothes and malnourished, the Turkish army was decimated by typhus and cholera, which killed more than 90,000 soldiers at the beginning of the following year. The survivors were taken prisoner.

Armenian provinces, now the theatre of a war, saw an inexorable Russian advance, accompanied by the first massacres of Armenians perpetrated by the withdrawing Turkish army. Armenians were accused by the Turks of being the main responsible for the Ottoman defeat. Soldiers soon whipped Muslim and Kurdish populations up against the Armenians. In February 1915 all Armenian militaries of the Ottoman Army were relieved of their duties, deprived of their weapons and put to penal service in groups of 50–100 men each. These groups were progressively isolated and their members executed. At the same time, passports were withdrawn to all the Armenian civilians while government officials were immediately discharged [4].

2.3.2   Reasons and Consummation of a Genocide

In November 1915, the Turkish government called the Djihad[4] taking advantage of the confusion generated by the withdrawal of the Turkish troops and the simultaneous advance of the French and English fleets into the Dardanelles, therefore handing Armenian civilians over to overexcited Muslim people [3].

Actually, deportations had already begun in April 1915 in Zeythun, far from the front, following successive phases of a carefully planned project. The pretext to the central authorities in Constantinople was provided by the resistance of the Armenian population, who saw their villages plundered and razed to the ground during the retreat of the Turkish troops: without food and ammunitions, the Armenians placed along Van Lake were rescued by the Russian army, led by Armenian volunteers. The population of that area was then saved from extermination and most of them decided to go to thicken the soviet ranks by enlisting in their army.

Soon after, Saturday 24 April 1915, all the Armenians intellectuals, journalists, politicians and priests (650 in total) of Constantinople were arrested and deported to be executed in the following months. To justify these captures, the Union and Progress Committee adduced the existence of an extended conspiracy, for which a sham trial was set up. The Armenian prisoners not yet deported were condemned to death and hanged. Since then, 24 April will be remembered as the date that officially opens the deportations and will be celebrated every year by all Armenians in the world as a memory of the genocide.

All the Armenians who were still in the eastern towns began to be deported: orders were given to Armenian families (announced or posted in every city) to collect their personal belongings and leave immediately. All remaining goods were seized with

---

[4]In the Islamic culture, this word has a wide range of meanings; in western countries, however, this word has been mostly understood as the Holy War against infidel and the armed tool for the dissemination of Islam. Djihad can be declared only by the maximum Muslim authority, traditionally the Caliph.

immediate effect, to be resold at ridiculously low prices or destroyed. Family heads were instantly and summarily executed, so that often the convoys of the deported were made of elders, women and children. On the Black Sea and the river Tiger many vessels transporting Armenian people were sunk; the Armenian provinces under the Turkish domain were systematically looted; everyone could act with total impunity: the only crime recognized by the Ottoman authorities was the protection offered to Armenian civilians. The high number of such episodes made it impossible for the Turkish authorities to keep them hidden to the world: thanks to the testimonies of ambassadors and missionaries, Turkey was pressed by the International Community to put an end to the massacre. This situation led the Turkish authorities to make official decrees, accusing the Armenians of collaboration with the Russian army, sabotage and terrorist actions: such charges were never proven. Turkish Armenia soon became scorched earth: of the 1.2 million Armenians who lived there, 300,000 reached the Caucasus thanks to the Russian occupation, 200,000 women and children were kidnapped and forced to convert to Islam, 50,000 managed to reach Aleppo, last meeting point for caravans. The remaining Armenians were eliminated, piled on the streets, hanged on telegraph poles or trees, thrown into the Euphrates' gorges, or left starving on mountain paths chased by Kurdish tribes. The Armenians living in western Turkey and those who were further away from the Russian front, not being an imminent threat for the Turkish army, were transferred in longlasting train journeys that had to be paid by themselves. As the tunnel across Aman and Taurus was not yet finished, the deportees had to walk along the mountain paths: many of them were then piled into makeshift camps where, undernourished, fell ill with typhoid and dysentery. Those who managed to arrive in Aleppo were moved to concentration camps set up in Hama, Homs, and nearby Damascus (Syria) [4].

In August 1916, Constantinople ordered the Armenian final resolution: when the deportation would be completed, those who were still in concentration camps had to be sent in the southern deserts of Syria and west of Mesopotamia, without water nor food. This order, which implied the massacre of further hundreds of people, eliminates any remaining doubt: the deportation aimed at the whole annihilation of the Armenian population. Many Armenians succeeded in escaping the massacre by reaching the Allied ships, or with the assistance of the Apostolic Nunciature or American missions: in Anipemza the Americans built an orphanage for the children who escaped the abominations of the deportation [5]. In total, considering also those who were able to reach the Russian territory, only 600,000 Armenians out of an estimated population of 2,100,000 were spared. This was the first genocide in the twentieth century. It was perpetrated by the Ottoman Empire in a systematic manner and relied on the concepts of nation and race; it was directed by the Young Turks, who unleashed the hatred of the Muslim populations against the Armenian gâvurof[5] Christian faith, thus hiding the massacre under the false mantle of a holy war [4] (Fig. 2.8).

---

[5]It is a word used by Turkish people to mean non-Muslim people.

Fig. 2.8 Children of Anipemza in the '40s (top) and '30s (bottom). (From private Anipemza's people photo books)

2.3.3 The Sovietisation of Armenia

The October Revolution caused the disintegration of the Russian army deployed on the trans-Caucasian front, making the Soviet troops withdraw into their native homeland. The confused and disorganized with-drawal offered Georgians and Tatars the opportunity to arm themselves at the expenses of the Russian army. On 18 December, the Russian army General Prjevalski, signed an armistice with the Turks, causing the respective armies to remain locked in their positions until the signing of a peace treaty. On the front, more than 500 km long, only the Armenian troops remained: just short of 20,000 people of the initial 500 hundred thousand.

The armistice was respected by the Turkish troops until the end of January; in February the Turkish army increased the guerrilla warfare quickly and accused Armenians troops, sometimes correctly, of atrocities against the Turkish population, thus giving them the justification to resume the fight. In front of a Turkish army much larger for soldiers and weapons, the Armenian troops were forced to retreat losing cities like Erzincan, Erzurum, up to Sarikamiş however managing to repatriate the Armenian population in Transcaucasia. On 22 April, the Mensheviks[6] within the Seim (Legislative Assembly of Transcaucasia) signed the declaration of independence of the Transcaucasia, organizing themselves in a Democratic Federative Republic completely separate from Russia [6]. It was then necessary to resume new negotiations and in Batum on May 11 the Turkish delegate formulated new demands claiming the Turkish Armenia in its integrity, including Anipemza, Kars, Ardahan, Batum, and part of the Yerevan provinces. Meanwhile the Turkish army, waiting for orders, deployed in the plains below the Ararat.

At this precise moment the Federative Republic of Transcaucasia turned out as a completely incoherent project: the Muslims of Dagestan wanted to join the Turks, while the Georgians, who firmly opposed the Turkish expansionism, received unexpected help from the Germans who advised them to withdraw from the federation, standing under their protection. On 26 May the independence of the Georgian Republic was declared and subsequently the Federative Republic of Transcaucasia fractured following its national lines. On 28 May the Republic of Azerbaijan was proclaimed: the Armenian Council soon realized that proclaiming the Armenian independence would be a catastrophe, since two thirds of its territory was occupied by enemy forces and more than half of its population consisted of refugees [7]. Surprisingly, however, the Turks recognized the new Armenian authorities and, on 4 June, proposed a new peace treaty: the new Armenian Republic was left with the cantons of Nor-Bayazet, part of eastern of Aleksandropol, Etchmiadzin, Sharur-Daralagiaz and Yerevan, which became the new capital.

Yerevan was a city with no economic or cultural importance, not connected to either the Caspian Sea or the Black Sea, but only to Tbilisi (known as Tiflis in Armenian); therefore, the newly created state had as its capital a city of wooden houses with no paved roads. From 1918 to 1919, Armenia was in a permanent deadlock with no means of transportation and with completely disrupted transport routes. The whole country was in a state of immeasurable misery: most of the population found itself living in ruins, with no medicines, food and clothing. During the winter of these years, typhus and cholera claimed a great number of victims, amounting to the 20%

---

[6]The Mensheviks (which in Russian means minority) were a faction of the Russian revolutionary movement that emerged in 1903 after a dispute between Lenin and Julius Martov, both members of the Russian Social Democratic Workers Party. The divisions between the two factions were of old origins, and had to do with pragmatic issues based on the history of the failed revolution of 1905, and on theoretical questions about class leadership, class alliances, and bourgeois democracy. The Bolsheviks (which in Russian means the majority) believed that the working class was to lead the revolution in alliance with the peasants. The Menshevik vision was that of a state more similar to Western democracies and of a gradual approach to socialism.

of the total population. Georgia in the north was falling under German control and the Turks were to the southwest [7].

Americans had forecast an aid plan with the distribution of food for the population through the American Committee for the Relief in the Near East (ACRNE): this aid had to pass, however, across Batum and Tbilisi and the Georgian government let go less than a third of it. Only at the end of May 1919 the United States of America were able to allocate to the Republic of Armenia financial aid amounting to twenty million dollars, ensuring some relief [7].

It was during this period that, despite inflation at record highs, the national government sketched a reorganization of the country: textile factories were renovated, the road system was rebuilt, the cognac and wine industries were nationalized. Armenian was adopted as the national language; the primary and secondary studies were made compulsory and free of charge. At Aleksandropol, the current Gyumri, the first Armenian university was opened: in October 1920 it was moved to Yerevan. Also, the first archaeological excavations were financed and opened [7].

At the diplomatic level, however, Armenia could not find good allies and this situation was clearly evident at the Peace Conference planned in Paris in 1919: the Armenian delegation claimed a very large territory that envisaged the annexation of the seven eastern districts and Cilicia to the already existing Armenian Republic, demanding that this new state was supported and as-sisted for twenty years by the League of Nations or any of the allied powers appointed as agent. They also demanded the implementation of reparations because of the war, the return of all forcibly converted Armenians, the trials against those responsible for the genocide. On the initiative of the Allied governments, a commission was created to conduct a general inquiry into the nations that were to be separated from the Ottoman Empire, giving the US a general mandate on the territories of Asia Minor. Major General Harbord, director of a military commission, went to Turkish Armenia and Transcaucasia and collected testimonies of the atrocities of 1915. These allowed him to attribute the responsibilities of the genocide to the government of the Young Turks. He noted that most people considered a good relationship with Russia the only way to obtain economic stability and external security. Gérard Dédéyan, in his text, asserts the Harbord report to be the source of the American Senate's refusal of the mandate on Armenia [4]. Indeed, the presence of the Russian army was necessary for the security of the Armenian boundaries and population; the problem was that Bolsheviks did never recognise the independence of the Armenian Republic and, in the meantime, diplomatic relations between the two countries increasingly deteriorated.

Stuck in the grip between the Turkish nationalism and Russia, still grappling with the aftermath of the Civil War, Armenians relied on a peace treaty to be concluded between the winners of World War I and the authorities of what remained of the Ottoman Empire [3].

In April 1920, in San Remo (Italy), Francesco Saverio Nitti, Lloyd George, and Clemenceau ratified the creation of an Armenian state with boundaries including part of the provinces of Trabzon, Van, Erzurum and Bitlis, while Cilicia, the regions of Maraş and Diyarbekir would be annexed again to Turkey.

At the Spa Conference, in the lounge of Sevres porcelain, a peace treaty between the allies and the government of Constantinople was finally signed. It recognized the sovereignty and independence of the Armenian state: this treaty was never ratified, however, because the kemalsti,[7] at the time very powerful in Turkey, did not recognize it as legitimate [3].

Vanished the dream of an independent Great Armenia, gradually abandoned by all Western countries' allies, the Republic soon became a land of conquest for Turks and Soviets.

The Bolsheviks initially tried a legal political annexation as already done in neighbouring Azerbaijan; the Armenian communist party though had not establish itself as a strong force yet and the attempt failed. In the south of Ararat, a Turkish attack, ordered by Kemal, with a 70,000-strong contingent caught unprepared both Armenians and Bolsheviks [3]. Armenia, exhausted by the events of the previous years, had enough strength to slow the Turkish army but not to prevent them to breach its defensive lines; Stalin then decided to take advantage of the situation and ordered the Soviet army to cross the boundaries of the state and headed towards Yerevan. The aim was to reach the capital before the Turkish armies could do it. The Red Army encountered no resistance either by the Armenian army or by the population who refused to take up arms against the Bolsheviks. The project succeeded and on 20 December 1920 the Russian Military Revolutionary Committee decreed that the laws of the USSR were also to be applied in Armenia.

The people welcomed with joy the news of the Sovietization: although this put an end to the independence, it also meant the return to a Russian protectorate with the pacification of the territories and a guaranteed continued vigilance on the Turkish border [6]. The sovietisation of the country though, soon proved to be brutal, despite the constant warnings of Lenin to deal with the Caucasian comrades with tact and respect. Finally, the Armenian population rebelled and overthrew the Soviet regime. On 8 March 1921, backed by the tacit support of the Turkish government, Armenia managed to restore its sovereignty over most of the territories annexed by the Soviets. This sovereignty, however, lasted just over a month, when a second wave of Soviet military returned to Yerevan hunting the political leaders of the revolt. It is at this point that a treaty between Moscow and Ankara on the geopolitics of Trans Caucasus was signed on 13 October 1921 in Kars.

It definitely closed the Armenian question [6].

The Soviet Union annexed almost entirely Georgia; Karabakh became an autonomous region attached to the Azerbaijan Soviet Socialist Republic, but it had to leave to Turkey 25,000 km^2 of territory along the administrative units of Kars, including all of the Ararat region and Ani. Anipemza remained Soviet territory located along the natural border marked by the Akhurian River [6].

A year later, on 13 November 1922 in Lausanne, a new international conference was held: the Turkish plenipotentiary Ismet Inöunü accused Western nations of being

---

[7]It is the name given to the supporters of the national liberation struggle of the Turkish people led by Marshal Mustafa Kemal Atatürk culminating in 1923 with the foundation of the modern Republic of Turkey and the end of the Ottoman Empire.

too meddling about question concerning the Turkish state and pointed the responsibility of the long-lasting Armenian question to the Tsarist Russia and the Armenian Insurrectionist Independent Movement.

He said the government and the Turkish people resorted to measures of repression or retaliation only after exhausting their patience [7].

To date, this is the version that many of the nations in the world have accepted as true and this allowed Turkey to free itself from any responsibility for the massacres, as well as avoid the payment of substantial damages. In the final report of the Committee, the "Armenian issue" was put on the Index of unexplained cases. The Armenian political situation will therefore remain the same until 21 September 1991, when a referendum restored the independence of the country, two months after the dissolution of the Soviet Union [4] (Fig. 2.9).

2.3.4   Anipemza: City Development, from an Orphans' Village to a Company Town

During the second decade of the twentieth century, after its annexation to the Soviet Union, the socialist buildup spread all over the country: private and public construction, canals, hydroelectric reservoirs, dykes, railways, reclaimed lands, mines, oil wells, kolkhoz and sovchoz[8] were all part of the five-year plan[9] [7] (Fig. 2.10). In 1924, four years after the start of the occupation by the Red Army, a plan for the rebuilding of Yerevan was carried out by the eminent Armenian architect A.I. Tamanyan; two years later Anipemza, one of the first industrial cities planned and completed during the soviet period, was founded. Tamanyan had allegedly a role in the urban planning of Anipemza, although his involvement has yet to be officially proved; undoubtedly, the city was erected by the inhabitants of the nearby village Zagha, many of whom were skilled workers. According to the testimonials of the older interviewed inhabitants of Anipemza, Zagha was a village who hosted the

---

[8]Collective and soviet farms owned by the state; they were a product of the collectivization of the land carried out by Stalin in 1929–30. The Kolkhoz was a sort of cooperative managed by an elective administration board and a president (the latter chosen by the party), that used collectively a state-owned land, with the possibility for individual farmers to own a parcel adjacent to the one owned by the state. The sovchoz, instead, worked as an industrial company. The farmers who were employed there, were state employees to all effects: the entire harvest was a state property, and the farmers received a regular pay.

[9]The five-year plans were introduced for the first time in the URSS under the guidance of Stalin during the years 1929–1933. The main body responsible for the economic five-year planning was the Gosplan, or the State Planning Committee. The first five-year soviet plan (1928–1932) fostered a great development of the heavy industry, while it damaged the production of consumer goods and the agricultural sector. The economic policy keystone during that time was the enforced land collectivization, according to which the kulaki properties were stolen from the institutions and transformed into kolchoz and sovchoz: these were forced to sell a part of the product to the state at a price set by the state itself in order to reallocate and feed a market whose incomes would be reinvested in heavy industrial activities.

Fig. 2.9   Soviet period Anipemza village signs

Fig. 2.10 Prisoners work at Belbaltlag, a Gulag camp for building the White Sea-Baltic Sea Canal. From the 1932 documentary film, Baltic to White Sea Water Way. Courtesy of the Central Russian Film and Photo Archive

orphan of 1915 genocide from some villages of the west Armenia (mainly Kars) and from Gyumri (Aleksandropol at that time) in 1916 to 1920. Americans took the population from orphanage of Leninakan, today known as the city of Gyumri, to work in mine.

In 1926 was approved a group of researchers to define what kind of construction materials possible to extract from Anipemza's area. Seven kilometres away to northeast, the research group evaluated subsequent types of construction supplies: lump pumice stone, construction materials and pumice stone for cement production. In the same year, by decision of the government mining work began starts in the quarry. First workers mined the materials by hand and only later by heavy machines. With the opening activities of the industrializing quarry, specialists were coming with their families from other nearby villages and cities too in order to work there. In 1932, due to the lack of a houses, they decided to build temporary settlement with minimal amenities for living.

The settlement expanded very quickly, and by 1932 they already had a seven-year primary school, which in 1949 became a high school. By that time, they had two grocery stores, one hardware store, hospital pharmacy canteen bakery, etc. In addition, some orphans who were moved from Greece during the Genocide, have moved to work in Anipemza in 1934. Starting from 1938, by decision of the government, the settlement was recognized as an urban village, which included residents of the neighbouring villages of Zagha, Abdul Rahman, Kharkov, Aikadzor, Bagrevand. Later to the 1950s workers from Spitak also joined the ranks. From the beginning of the 1960s due to lack of drinking water and basic facilities and because of industrial development in other regions huge number of workers moved to Abovian, Hrazdan, Charentsavan, Arzni and other developing cities. Since the 1970s, the Combined production has been expanded and new divisions have been opened as crushing screening department and department for producing dust of andesite (basalt). After the 1988 earthquake the division of plant were turned into cooperative production: facing tiles, which is still working today. Between 1991–93, after the Azerbaijan attack to Armenia and because of the increase in the cost of railways, the exportation quitted and the plant stopped. In 2003 the quarry turned into Open Joint Stock Company. Today part of it is privatized and they produced tuff tiles for covering.[10] Today (2018) only 10 people work in the quarry, only a few buildings of the wide tuff and pumice Anipemza's mine, remained. The actual mine owner demolished progressively almost all industrial buildings and warehouses to pay less taxes.

The inhabitants recounted about the prosperous period of the village, when there were different types of manufactures there. Particularly they mentioned the stone mine which was exploited well and produced stone blocks were used in construction works of many outstanding buildings in Yerevan and Gyumri, such as the Yerevan State University designed by architect Alexander Tamanyan[11] (1878–1936). The salary of mineworkers was very high; people were working on 3 shifts. There was a lot of work. They should load 100 wagons for Kaspi, Ararat, and other sites. After the end of World War II, the city was enlarged with the construction of nineteen residential buildings (120 new apartments); the school, the library, the house of culture, the people canteen, the hospital, shops, and other service buildings were also built during that period. The garden, visible from the main street, is enclosed by fence and nowadays is used as a private vegetable garden and orchard. There are three shops which serve Gyumri too. The interviewed people said that during the Soviet period,[12] life in Anipemza was perfect, they were rich, people had job, also the residents of other villages were daily commuting to the village, taking advantage of the today missing Anipemza Train Station (Fig. 2.11).

During the visits in site we discovered that Anipemza was also both a forced labour settlement camp for Armenian dissidents during the Soviet period. Political prisoners were staying in the two-storey building in front of the former public garden (Fig. 2.12) and each room accommodated up to four dissidents: they could not receive visits from people who did not have special permits and the entire building was watched day and night by armed troops. This solution did not require the use of

---

[10] Specific data about Anipemza's quarry reported in this paragraph has been collected by Marco Germi, who interviewed Tigran Kakolojan in 2018. He was chief power engineer in the quarry since 1987, then became leader of crushing-screening departed. Now he's a schoolteacher.

[11] Tamanian or Tamanyan, according to different transliterations.

[12] Armenia has been independent from Soviet Union since 1991.

fences or other architectural barriers. The entire village did never have bare walls or barbed wire fences, but had a toll station located on the only access artery of the village—a common solution for all border villages; Anipemza, however, until 1985, could only be accessed with a special pass to be asked to the authorities.

The village is served by the railway linking Gyumri to Yerevan, but the nearest railway station is Aniavan, a village about 5 km away; a railway track, still visible, reaches the quarry and is used only to connect the country's production area with the other two major cities. From Gyumri to the south, paralleling the border with Turkey, there is a road that crosses the only rail link between Armenia and Turkey: this rail link was opened in 1898 to provide a connection between Tiflis (nowadays Tbilisi in Georgia) and Kars (now a Turkish city) at a time when both cities were geographically embedded within the Russian territory. After the Kars treaty of 1921, until the collapse of the Soviet Union, there was a single train per week that still moved along that route, abolished with the closure of the border in 1992. The area is of strategic importance because it marks the border between the countries of the former Warsaw Pact and those of NATO and it is still very tangible due to the continued presence of Russian soldiers. Designed by the Soviets as an industrial town, Anipemza did never invest in pastoralism or in agriculture also because of the scarcity of water. Still today the city lacks a central system for the supply of drinking water; the water needed for the quarry and for irrigation comes from the neighbouring communities or is pumpedout from the Akhurian river which marks the current border with Turkey. Despite the various changes that Anipemza has suffered over time, it still appears today as an industrial colony of utopian socialist ideals developed in the second half of the nineteenth century. Such ideals led to the creation of the garden cities of Ebenezer Howard, which is very similar to areas that we can still visit today in England, Italy and the rest of Europe. It is so unique in Armenia that it deserves a strong interest, not only from national governments, for its historical, sociological and cultural significance but also for the abovementioned urbanistic and architectural features.

The question is: Who designed the village? Anipemza is not a spontaneous settlement as many in Armenia's territory. The urbanistic implant suggests that it was designed by a great architect. The most important Armenian architect of that time was, as already mentioned Alexander Tamanyan, who, in 1919, moved to Yerevan and was appointed as President of the Preservation of Antiques and the Department of Main Architecture.

Back from Russia he stayed for a short period in Tabriz, Iran, then moved to Armenia, in 1923 designed the new Yerevan town and from 1925 to 1932 planned towns and settlements in all the Country as Leninakan (today Gyumri), Echmiadzin, Arabkir (a district of Yerevan), Loukashin, Oktemberyan (today Metsamor), Step'anakert (today in the Republic of Artsakh), Nor-Bayazet (then Kamo and today Gavar), Akhta-Akhpara (today Hrazdan), Nubarashen (district of Yerevan today Sovetashen). All these plans had precise compositional scheme and a clear idea inherently the garden-town principle. Unfortunately, no documents about Anipemza have been found the consultation of Tamanyan Historical archive in Yerevan [8]. It seems that Alexander Tamanyan and another famous Armenian architect, Toros Toramanyan, restored the cathedral of Yereruyk. So, for sure, Tamanyan already known

Fig. 2.11 Historical photo (late '70) of the railway station of Anipemza. (From private Anipemza's people photo books)

Fig. 2.12 A residential building transformed into forced residence for political dissidents during the Soviet period of Armenia

Anipemza and someone (many of the elder of Anipemza) say that Tamanyan was the designer of Anipemza even if still there are not archives documentation evidences support this hypothesis.

Researches are still in progress. However, some similar evidences suggest his style. Maybe used by a direct or indirect pupil of him. For example, a Tamanyan's 1925 project (only one year before Anipemza) of "New Arabkir" town (now a district at north of Yerevan centre) discussed by Hurutynyan shows the floorplan and elevation of a double residential unit [9] like the prospectus and plans used in Anipemza only one year later (Fig. 2.13). Besides, the plan of Anipemza suggest also the typical composition used for the Soviet Gulags with a central axis and dwellings as well as facilities (canteen, offices, House of Culture, and so on) arranged regularly on the both sides of the main road (Fig. 2.14).

2.4 Anipemza Today: Walking Through the City

Anipemza can be reached through a single access road from the Yereruyk basilica: the still paved avenue is now in a bad state of conservation, but still has on its sides flower beds, trees and cast iron lighting systems with refined decorative details. It is clearly felt that the village is divided into two, thanks to the architectural choice of constructing on the left side mainly two-storey buildings, and on the right side single floor buildings: the only exceptions are the building that was used to host the quarry management offices until its closure in 1994 (two-storey) and another, more recent (three-storey) built when the head of the Soviet was Leonid Il'ič Brežnev. Building materials are mainly tuff, extracted from the quarry, and wood, with the exception of some portions of buildings which were made of reinforced concrete, in the early 1960s [9].

When the chromium of the tuff blocks becomes orange in colour, it is certain that this material has been extracted from Anipemza, because Armenia has the only quarry with such characteristics. It is by no means certain that the tuff blocks used to build the Church of St. Grigor in Yerevan come from here. The use of bricks is very rare, but not for the production of fumes and chimneys; for roofing the most used material is iron (metal sheets).

Today Anipemza has less than 500 inhabitants, most of whom commute to and from Yerevan to work, making the village inhabited by elders and children throughout most of the day, for most of the week. Lately, when reaching legal age, most youngsters leave Anipemza to seek work elsewhere so as to make a living. This is making the problem of depopulation of the village more and more evident.

During the Soviet period, according to the interviews made in place with the elders, the situation was very different: the village was densely populated and totally self-sufficient.

The first buildings to be built were the two-storey buildings on the left, with carving walls up to one meter thick. One-storey buildings were only built later and elders in the resort still name them by the name of the different owners ("the House of

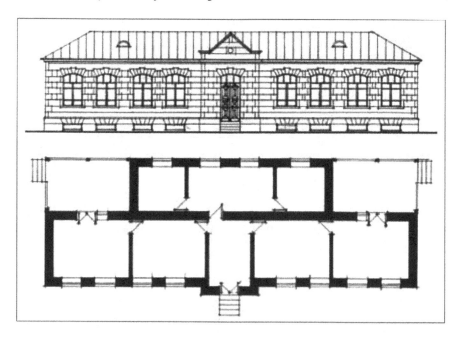

Fig. 2.13 A Tamanian's 1925 project (only one year before Anipemza) of "New Arabkir" town (now a district at north of Yerevan centre) discussed by Harutynyan in her book

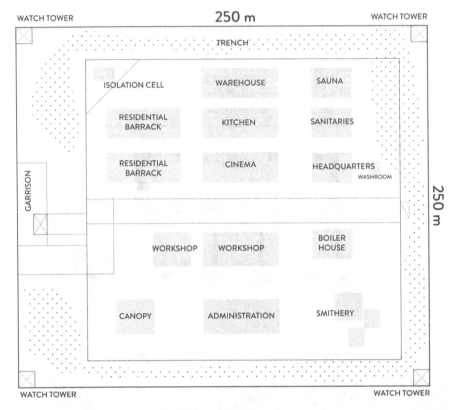

Fig. 2.14 Soviet gulag scheme with many similarities with the planimetric scheme of Anipemza. Plan realized after Prison Plan of Perm-36, original sketch made by Lett Gunar Astra (in Latvian). Courtesy of the Gulag Museum at Perm-36[13]

Gayan", "the House of Aramik", etc.). The various one-storey residential buildings are categorized according to the number of rooms ranging from one-room to four-room apartments where more than one family can be found (N 30 in the Fig. 2.15) [9].

Originally, the apartments were all designed without kitchen or bathroom, as the inhabitants used the public baths and the people canteen made available by the state-owned company that conducted the excavation works in the quarry. The village was also equipped with a bakery, which is now reduced to a cumulus of rubble where it is not uncommon to see grazing sheeps and pecking hens. Unfortunately, with the fall of the Berlin Wall and the consequent collapse of the USSR, the village has gradually begun to decline, ending up in the oblivion of the Armenian government apparatus; only recently recovery plans for border villages are being examined.

---

[13]From the online exhibit "Gulag: many days, many lives". http://gulaghistory.org/nps/onlineexhibit/stalin/perm36.php.

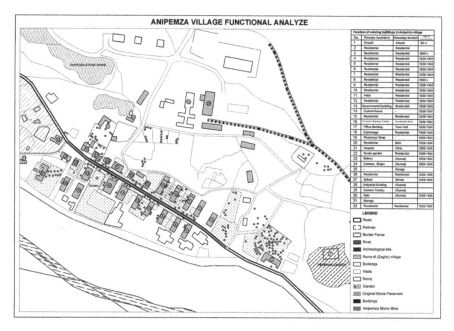

Fig. 2.15   General overview of the village: buildings, functions and construction year [9]

2.4.1   Problems

The mayor of the village, Harutyun Tarlanyan, interviewed several times, is trying to focus the attention of the national authorities on Anipemza, bearing in mind how disastrous the situation has been so far: in a little over a century the population has passed from 2,359 inhabitants (in 1873) to 392 (in 2013), of which just a few less than 300 has is resident there. Males account for 49% of the total population. Children account for 32%. 53% is workingage population while the remaining 15% are aged over 65 (Fig. 2.15).

By this way—the mayor explained—schools will have to close, social conditions of the inhabitants gradually deteriorate, since they lack the means of subsistence, there are no agricultural fields to cultivate, there is no development and there are no future prospects.

Some of the residents are working seasonally in nearby stone mines. Many of them have left the village long time ago and others are still leaving. People interviewed by us have told: "The village is already empty. Young residents are leaving. The ones who remain are working in the mines. If they do not go away, they can't live. It was better before. Now young people are going away. Families are also leaving the village because of their children. Before it was not a village, it was an amazing borough". Those who have not gone yet have no financial means to do so.

The wire border passes just near the school. On the backside there are viewpoints of frontiers. In an interview[14] the journalist asked to the teacher, Tigran Hakobyan, if the situation of border dose straiten them. He answered: "Honestly no. I was born and grew up here. For me it is regular. Now the world has developed. Only this border of 360 km and the wires. Also, we can say to the pupils that this separating border divides us from our historical land. It is not a border between two countries, but a border between our past and present". Tarlanyan voiced as a concern: "The number of the population drastically went down in the last 20 years. The number of school age children has been declining and the community has registered more elderly in the total number of the population. At present 10–15% of working age men migrate for earning bread. This tendency is high among young families. Only 20% of the population is employed". One of the biggest problems is water supply; citizens are forced to make do, storing as much rain as possible through drainage pipes connected to common tanks. This water is treated twice a week, but to get some drinking water, residents must leave the village. They pay 200 drams for 20 litres. The community manager, in the interview, said that they have requested for final solution of water problem but certain date for that has not been announced. They have organized the supplying water system, but it is only irrigative water not drinkable. Drinkable water is imported with cars, irrigation water is not delivered to the area and people use rainwater preserved in bins. There is no canalization system in any of the buildings, no toilet, no bath; each of them tries to create the required facilities on their own.

By visiting the private gardens in the old public garden of Anipemza, it is well visible how everyone uses tanks and bins to store rainwater for private irrigation since there are no hydraulic systems. This can be seen in every apartment where everyone tries to ingeniously build structures to satisfy their needs.

The paucity of water is the basis of the choice to invest relatively little in sheep farming: animals need a large amount of water, and its cost would not be sustainable for their owners. With the little they can get from their own gardens and their animals, they try to satisfy their own needs and those of their family members, advancing little or nothing for sale to third parties. Many people prepare bread in their houses and all the remaining foodstuffs are imported from Gymuri.

As for gas supply, there is a pipeline which works perfectly: the problem is the money to pay for the heating. The Armenian Government has established an ordinary maintenance project to make some targeted interventions, due, on the majority, to roofing problems. The roofs made during the Soviet era lasted a little less than a century before beginning to highlight problems of water infiltration, but all the roofing, almost half of the existing ones, which were restored, showed strong deficiencies within a few months. Actually because of strong winds, whole portions of roofs fled. In many cases it could be ascertained that inhabitants of some buildings do not provide for repairs of the damage on the roofs waiting for a government intervention, despite the harsh climate of the region (very rigid winters with heavy snowfall and extremely arid summers interspersed by strong hailstorms with annual

---

[14]http.//www.azatutyun.am/media/video/25139611.html.

rainfall of 500–600 mm). Another risk factor for the village, the most dangerous, is its position in a seismic area.

In short, the most important problems to solve are:

- find new job opportunities for Anipemza's inhabitants;
- repair of roads;
- renovation and improvement of drinkable water pipelines;
- renovation of school furniture;
- realize heating devices and gas delivery;
- sales of agricultural products.

Buildings need improvements in a conservative way.

So, the community issues are:

- school furniture: 3 million drams required, or € 5,500[15];
- 60 km of drinkable water pipeline construction: 600 million drams required, or € 1,100,000;
- 5 km of pipeline construction: 40 million drams required, or € 73,100;
- 7 km asphalt road: 280 million drams required, or € 511,800;
- repair and furnishing of health centre: 30 million drams required, or € 30,900;
- repair and furnishing of cultural house: 30 million drams required, or € 30,900;
- repair of elementary school building: 15 million drams required, or € 28,000;

The solution of main problems of the community requires around 998 million drams, approx. € 1,900,000 [10]. Nevertheless, Anipemza is still today a very intriguing place, whose history enriched it with details so peculiar to make it a unique case in Armenia.

The authors of this paper maintain that it would be enough to invest a limited amount of economic resources to improve the current situation: a good start would be the cultural interest that this village can surely demonstrate to its visitors. Anipemza should be inserted into a common network of tourist routes (Armenian churches and fortres-ses), as a unique witness of the most significant historic events of the twentieth century recognisable in its architectonic heritage.

References

1. Donabédian P (2014) Ereruyk: nouvelles données sur l'histoire du site et de la basilique. Mélanges Jean-Pierre Mahé, Travaux et Mémoires 18, Paris, pp 241–284
2. Alpago Novello A (1995) The Armenians: 2000 years of art and architecture, booking international
3. Akçam T (2005) Nazionalismo turco e genocidio armeno: dall'impero ottomano alla Repubblica. Italian edition by Antonia Arslan, Guerini, Milano
4. Cemal H (2015) 1915: Genocidio armeno. Guerini e associati, Milano
5. Bobelian M (2009) Children of Armenia. A forgotten genocide and the Century struggle for justice. Simon & Shuster, New York

---

[15]Current (2018) exchange rate has been used for all approximated calculations in the chapter.

6. Payaslian S (2011) The political economy of human rights in Armenia. The authoritarism and democracy in a former Soviet Republic. Tauris & Co Ltd., London
7. Dédéyan G (ed) (2002) Storia degli armeni. Italian edition by Antonia Arslan and Boghos Levon Zekiyan, Guerini, Milano
8. Hurutynyan E (2009) The durable handwriting of Armenian, Yerevan
9. Augelli F, Khachatourian Saradehi A, Khachatourian Saradehi L (2015) Anipemza: from genocide orphans' village to workers village. First proposals for conservation, valorisation and improvement of an interesting architectural settlement example and of a rich history site in Armenia, Scientific Papers of NUACA IV(59):14–28. ISSN 1829-4200, Yerevan
10. Tarlanyan H (2009) Program Anipemza rural community 2013–2016 social-economic development. Anipemza community manager, report. Anipemza

Chapter 3
Characteristics, Materials and Decay Analysis of Anipemza Buildings

Paola Bertò and Alessandro Marcone

Abstract The perfect knowledge of historical buildings is fundamental in the conservation project: first of all, building characteristics, dimensions, technologies and materials of Anipemza have been recognised and listed; after that, decay phenomena are deeply analysed to recognise pathologies acting on the materials and structures in order to understand causes and find solutions.

Keywords Survey · Traditional materials · Tuff stone · Basalt stone · Timber · Lime plaster · Typology · Construction · Abandonment · Deterioration · Lack of maintenance

3.1 Characteristics of Buildings

The perfect knowledge of the object is essential in the definition of a conservation action: to be successful, the project must be based on exact data, supported, if needed, by chemical and physical analysis as well as in site tests. This will avoid then the risk of inadequate, when not counter-productive, interventions. A typological analysis and an accurate geometric survey seem to be the more logical steps to find information [1, 6].

In Anipemza different kinds of building typologies can be identified: as already described in Chap. 2, it seems likely probably that the urban masterplan was designed, between 1926 and 1936, by an important architect (Fig. 3.1). Residential buildings, common spaces and facilities stretches out along a central axis, decorated with flower beds, rows of trees and a cast iron lighting system. A public garden and sporting facilities were present in the past: now they are abandoned or re-used for private purposes. Public building (i.e. Dining Hall, House of Culture, Infirmary…) were also, probably, built at the same time (Fig. 3.2).

During the first village expansion two-storey buildings were erected, on the south-west side of the main road using local building materials. At a later time also one storey buildings were built on the north-east side. The construction system was based on load—bearing walls (perimetral and spine walls) up to one meter thick, realised with the tuff coming from the local quarry: stone was squared off in ashlar blocks and laid in horizontal courses, joined by lime mortar. Some blocks were perfectly worked

F. Augelli et al., *Preservation and Reuse Design for Fragile Territories' Settlements*, SpringerBriefs in Applied Sciences and Technology, https://doi.org/10.1007/978-3-030-45497-5_3

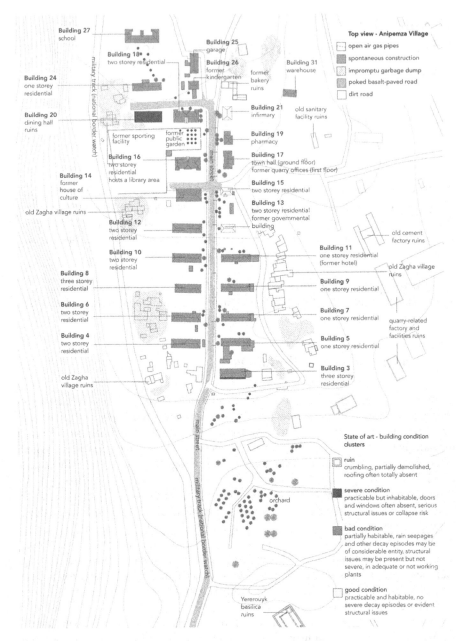

Fig. 3.1 General overview of the village: the buildings' typologies, the general conditions and the functions

Fig. 3.2 The former public garden, now used as private orchard, between building 16 and 18

and moulded on crowning, doors and windows, as opposed to blocks intended to be plastered, who were unrefined and showed working marks.

Roofs were realised with timber elements, avoiding the use of trusses, a highly recommended solution in seismic areas.

One-storey building on the north-east (right side coming from the settlement entrance) of the axis have mainly residential function: they have rectangular layout and no decorations on windows and on doors. Original volumes have been often modified as a result of the edification of new external construction on private parts (generally toilets) (Fig. 3.1).

The House of Culture, the Dining Hall, the Pharmacy and the Infirmary are also one-storey buildings: they all have a simple geometric plan. The first has a little protruding volume in correspondence with the entrance. The second has very large framed windows and a moulded crowning. The third building is smaller than the others, has an added volume on the side, and framed windows and doors, just as the Infirmary (Fig. 3.1). Surfaces of all buildings have been plastered or painted directly on tuff stone.

Two-storeys residential buildings can be divided into three typologies named below A, B and C.

Type A buildings (16 and 18) show a particular organisation of the geometrical plan: on a rectangular basis, the central and the perimetral parts stick out of the back of the building. This volumetric solution, combined with the presence of stony frames on the protruding parts, gives a dynamic rhythm to the whole structure (Figs. 3.3, 3.4, 3.5, 3.6, 3.7, 3.8, 3.9, 3.10, 3.11 and 3.12).

Fig. 3.3 The former sporting facility, East of building 20

Fig. 3.4 View of the one-storey residential building 9

Fig. 3.5 View of the house of culture, building 14

Fig. 3.6 View of the dining hall, building 20

Type C buildings have rectangular layout and they are longer than others. Original volumes had loggias in the middle and on the sides, but some of them were over time closed by inhabitants (Fig. 3.1). Windows are simply framed, whereas doors are not decorated.

Fig. 3.7 View of the pharmacy, building 19

Fig. 3.8 View of the two-storey residential building 18—type A

Building 17, whose ground floor hosts the town hall, is also a two-storey building: quarry offices occupy the first floor. Its volumetric system is similar to type A buildings, only dimensions are different.

In the 1960s two three storey buildings were erected to complete the current urban asset. One of them was aligned with the two-storey residential buildings on

Fig. 3.9 View of the two-storey residential building 15—type B

Fig. 3.10 View of the two-storey residential building 6—type C

Fig. 3.11 View of building 17

Fig. 3.12 View of the 1960s three-storey building 8

the left; the other, with its poor architecture (concrete balcony) and now badly deteriorated, was built directly facing the archaeological site of Yererouyk. Because of
its prominent position, this building represents the first approach to the village for

Fig. 3.13 View of the 1960s three-storey building 3, facing the archaeological site

a tourist visiting the basilica, hiding the high architectural and urbanistic quality of the settlement (Fig. 3.13).

The geometrical representation of the physical state of the village's buildings has been one of the first concerns of the present work.

The access to previous surveys' materials regarding the village was not available in the first instance, exception made for the products of the works conducted from Professor Francesco Augelli, Arin Khachatourian Saredehi and Lousineh Khachatourian Saredehi [2] already mentioned in Chapter One, and a collection of cadastral plans and records that were offered by the Anipemza Municipality. The first operation conducted was to reorder and understand these materials in order to have a general insight on the different buildings' typologies and the changes that occurred in the more recent times, transforming, sometimes radically, the original plans according to the people's needs. This operation was performed in studio, before the majority of the people involved in this publication got significantly or systematically in touch with the village's reality.[1]

Once on-site the preliminary survey's actions consisted in producing eidotypes[2], at the urban, architectural and interior scales (Fig. 3.14). This enabled a more conscious

---

[1] As previously mentioned in Chapter One, some of the authors of this book had the opportunity for a significant but time-limited visit to the village during a Politecnico di Milano's Architectural and Landscape Heritage Specialization School Armenia Study Tour in 2014.

[2] Topographic or architectural proportional sketches obtained through direct observation. They are used as a base to store dimensions' notes, as a tool to obtain a first reading of the proportions and the spatial configurations and as a base for further investigations and design.

Fig. 3.14 The on-site eidotyping operations: the creation of a sketch obtained from direct observation was used as a tool of understanding as well as a base for measuring's notes and further drawings' elaboration

and even time-saving survey design and measurement's operations as could give a preliminary yet consistent idea of the proportions and the formal relationships of the spatial configurations' elements (Fig. 3.1) [2].

This fact was even more important considering the limited time and resources available at that stage. For this reason, considering the urban scale, only general dimensioning was taken, using simple tools like metric ribs.

For what concerns the architectural and interior scale, only five buildings[3] were actually measured, those were selected taking into account the possibility to access at least the 50% of the building's interior spaces and considering a very preliminary strategic feasibility study that highlighted the potential case-studies for the Adaptive Reuse Design described in detail in Chap. 5 of this book. At this scale folding rules and portable laser meters were used.

For what concerns the outcomes of the materials collected on-site, as we could rely on more time and resources once back in Italy, even if we were conscious of the limits and the tolerances due to the tools and the accessibility problems we had during the survey, we anyway decided to use an hBIM[4] software that could enable us to create a complete 3D model of the village [3], from urban to the interior design scale

---

[3]Building 5, 14, 20 and 18. Building 3 was only measured on the outside due to accessibility issues. A 3D model from capture in motion was the only available solution to survey the South/Eastern façade's loggia, that will be further described in Chap. 5.

[4]Heritage Building Information Modeling.

of the considered buildings, that has proven an appreciable base for the production of both the Conservation and Adaptive Reuse Design materials.

One of the most difficult part was to develop a satisfactory model of the terrain, as no drone was available to us in Armenia at that time,[5] but before leaving from Armenia we could finally access a 1:5000 cartography of the area that provided us information about the contour lines on which we based the territorial virtual model and we could check and compare the information we collected on-site. We also double checked the result with the available satellite images, with all due approximation considerations concerning this data source.

The geometric survey of the buildings and areas represents the basis of the process of knowledge: all the physical peculiarities of the architectonic object must be inserted in it. However, in an isolated village, it is extremely likely that materials and decay are the same on all buildings: the analysis can therefore be conducted on an urban scale, thus highlighting common features and problems of the buildings. The use of technical sheets, in this type of investigation, seems to be a good strategy.

3.2 Materials

The first step of the analysis is the identification of materials which compose the buildings; different materials are recognised, inserted on the geometric survey base (Fig. 3.15) and finally described in detail through the annexed technical sheets. A good knowledge of materials is basilar and any related information can be helpful, when planning the intervention.

Every technical sheet shows the details of the single component: photos are provided and physical characteristic of the material are accurately reported in the *description* paragraph. The additional paragraph—*annotations*—informs about use and function of the material, highlighting at the same time any problematical connected situation [4].

Two sheets describing tuff stone (M01—Fig. 3.16) and wood (M07—Fig. 3.17) are showed, as example of many sheets done, to illustrate the analysis and record method.

The strict connection between material, decay and best practice technical sheets permits to have a global view of the analysis on the single data sheet.

Remaining main building materials of Anipemza are in short described below:

- *Asbestos tiles*—authentic asbestos components of the covering systems; rhomboidal–shaped and greyish; affixed on the wooden structure, they were later covered or replaced, when damaged, with iron tiles and corrugated iron panels.
- *Cement plaster and cement mortar*—realised with water, cement binder and aggregate; colour is grey/light grey; from very fine to fine crushed rock aggregate

---

[5]From drone photos we could have developed a Capture from Motion model of the whole surroundings.

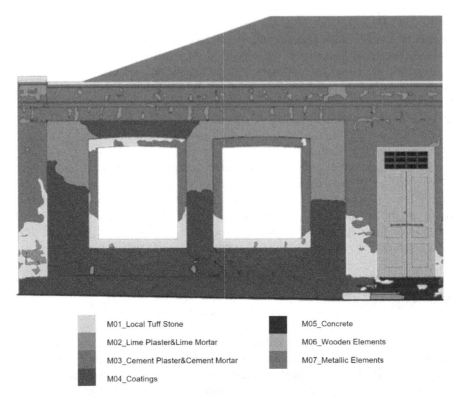

	M01_Local Tuff Stone		M05_Concrete
	M02_Lime Plaster&Lime Mortar		M06_Wooden Elements
	M03_Cement Plaster&Cement Mortar		M07_Metallic Elements
	M04_Coatings		

Fig. 3.15 A detail extracted from one of the material maps

(plaster) and medium crushed rock aggregate (mortar); colour of aggregate, most likely available locally, goes from light pink to black.

Cement plaster has often been improperly used to replace original lime plasters. Cement mortar has been used to repoint joints and to fix new building elements (i.e. aluminium windows).

- *Coatings*—on natural stone, plasters, wooden doors and windows, metallic elements; colour goes from white to yellow, and from light pink to green and light blue; used as protective or decorative layer, they are mostly peeling off.
- *Concrete*—pillars, beams, staircase realised with water, cement binder and aggregate; colour is grey; coarse crushed rock aggregate; colour of aggregate, most likely available locally, goes from light pink to black; reinforcement bars with increased adherence are present, where needed. Concrete is not a traditional material building in the architecture of Anipemza, except for the constructions of the soviet period: it was mainly used during restoration works ('50s).
- *Glass and ceramic elements*—bricks are rarely used in Anipemza; actually, only some damaged chimneys and roof structures are made of handcrafted bricks.
  Original windows had glass panels, most of which were broken and were somewhere replaced with plastic, iron or wooden sheets.

data sheet M01

LOCAL TUFF STONE

description

Tuff stone; colour goes from light pink to dark grey, depending on the seam; the stone was cut, scabbled (foundation blocks), worked until squared in ashlar blocks (walls), sometimes moulded (crownings, doors and windows...); many blocks show working marks; ashlar blocks lay in horizontal courses; joining material is lime mortar, frequently refluent.

annotations

Extracted from the local querry, the tuff stone is the most used building material in Anipernza (except for some concrete buildings from the 50's).
Facing blocks are perfectly squared and do not show working marks, as opposed to those who had been covered with plaster.

decay data connections

D01 blistering	D05 moist area	D09 film	D13 gap
D02 fracture	D06 staining	D10 graffiti	D14 moss
D03 craquele	D07 disintegration	D11 lichen	D15 plant
D04 deposit	D08 efflorescence	D12 missing part	D16 soiling

technical sheets connections

TS01A paving - restoration	TS06c cleaning - microorganisms
TS01B paving - creation	TS07 replacing of a disaggregated plaster
TS02 collection of garbage	TS08 repositioning of missing stone elements
TS03 restoration of a covering system	TS09 repointing of joints
TS04 creation of a rainwater collect system	TS10 restoration of an iron element
TS05 reorganisation of the electrical system	TS11 restoration of a historic iron element
TS06A cleaning - incoherent deposit	TS12 check of a wooden structural element
TS06B cleaning - efflorescences	TS13 restoration of a wooden door - window

Fig. 3.16 Data sheet M01

description

In the traditional architecture of Anipernza wood is the second builfing material. Ceilings, dividing walls, roof structures, sometimes floorings, balconies, cornices and original windows and doors are made of wood. The closing surfaces of ceilings and dividing walls are realised with wooden planks and battens, sometimes arranged on a regular grid and plastered.

annotations

Wooden doors and windows have been often replaced with aluminium ones in residential buildings.
Roofs are not realized with truss structure and therefore are not optimal structures for seismic areas.

decay data connections

D01 blistering	D05 moist area	D09 film	D13 gap
D02 fracture	**D06 staining**	D10 graffiti	D14 moss
D03 craquele	D07 disintegration	D11 lichen	D15 plant
D04 deposit	D08 efflorescence	**D12 missing part**	**D16 soiling**

technical sheets connections

TS01A paving - restoration	TS06c cleaning - microorganisms
TS01B paving - creation	TS07 replacing of a disaggregated plaster
TS02 collection of garbage	TS08 repositioning of missing stone elements
TS03 restoration of a covering system	TS09 repointing of joints
TS04 creation of a rainwater collect system	TS10 restoration of an iron element
TS05 reorganisation of the electrical system	TS11 restoration of a historic iron element
TS06A cleaning - incoherent deposit	**TS12 check of a wooden structural element**
TS06B cleaning - efflorescences	**TS13 restoration of a wooden door-window**

Fig. 3.17 Data sheet M07

Some original tiles floorings are present. More recent ones are generally realised with new tiles.

- *Lime plaster and lime mortar*—realised with water, lime binder and aggregate; colour goes from light pink to dove grey, depending on the colour of the sand; fine crushed rock aggregate (plaster) and medium to coarse crushed rock aggregate (mortar); colour of aggregate, most likely available locally, goes from light pink to black. Plasters are generally finished with coating layers.
- *Metal elements*—roofs, gutters, chimneys, covering systems, light poles are metallic, generally galvanised iron; some original cast iron elements are also present, like light poles and gutter downsprouts.
  Lacking doors and windows have been replaced with metal sheets or iron grates, sometimes varnished.

Aluminium windows were used to replace wooden ones in residential buildings.

3.2.1 Decay

The step that follows the identification of materials and techniques is the critical interpretation of their alterations and deteriorations causes and effects: the decay represents the loss of those chemical, physical, mechanical and formal peculiarities which characterized the materials during the time. The degeneration is caused by different factors, such as material peculiarities, installation, environmental conditions, lack of maintenance… therefore it can be asserted that not always a certain kind of decay is linked directly to only one factor. Additionally, it is not generally possible to know exactly the environmental conditions to which the material was exposed. Notwithstanding the above-mentioned difficulties of the decay analysis, this is the pivotal phase of the project: the understanding of the mechanisms at its origin actually allows to define the kind of interventions necessary to maintain the object guaranteeing the most successful and longstanding result [5].

For the natural and artificial stone materials analysis, the ICOMOS "Illustrated glossary on stone deterioration patterns" was used, but definitions were extended to all materials. Technical data sheets were created to provide photos and describe pathologies, standardizing the analysis with the previous one [1, 6].

According to the ICOMOS glossary [6], every degradation pattern belongs to one of the following families: crack and deformation, detachment, features induced by material loss, discolouration and deposit and biological colonization. The formal definition on the phenomenon is reported on the sheets and, additionally, hypothesis on the possible causes of deterioration are provided.

The compresence of various pathologies on the same surface is not uncommon: decay phenomena are often connected one another, ergo the analysis has to be exact and global at the same time, to understand cause and concurrent causes. Punctual interventions are not useful, if the global problematic has not been identified.

Main widespread phenomena—discolouration/moist area, disintegration, efflorescence/subflorescence—are shown below (D05—Fig. 3.18, and D08—Fig. 3.19) as example of technical sheet done.

Additional local decay phenomena and hypothesis on the causes are shortly analysed below.

- *Blistering* (family *detachment*)—caused by crystallization of the soluble salts under the superficial layer. Generally blistering precedes the localized loss of the superficial layer and is accompanied by lacunae. This kind of problem has been noticed diffusely on the bases of buildings, where rising damp causes subflorescences.

- *Fracture (*fam. *detachment)*—caused by earthquake vibrations, static problems, removal of the foundation earth and consequent decrease of the static balance, ground subsidence. The thickness of walls is not sufficient to guarantee a good structural dynamic behaviour of the structure: the visible system of cracks says that walls on the edges are not good connected one another, whereas a box-type behaviour in a seismic area would be desirable. More after, the no-truss roof structure transmits horizontal forces to perimetral walls, amplifying the reciprocal detachment. The problem concerns mainly older buildings, where ashlars are not good connected on the edges.

- *Craquele/Star crack*—concerning almost all plasters and mortars of the village, depends on earthquake vibrations, weathering, material shrinkage, rusting of iron elements in the walls.

- *Deposit*—lacking adhesion to the surface, the deposit of disaggregated materials (soot, dust, remains of conservation materials…) does not generally pose a threat to the conservation of the historic building. It concerns all buildings and surfaces.

- *Staining*—the presence of metal oxides, coming from an object corroding and driven by water onto the surface of the underlying elements, causes an aesthetical problem, because sometimes they cannot be removed from the stone surface. Wood also, after permanent weathering exposure, show discolouration, becoming greyish. Staining concerns generally exposed elements.

- *Film*—detachment generally provoked by weathering, thermal stresses, physical incompatibility between different materials. The term has been used to indicate a real detaching coating layer, formerly applied on stone and wood surface.

- *Graffiti*—generally resulting from vandalism acts. When graffities have particular value, e.g. historic/cultural value, should be preserved.

- *Missing part* and *Gap*—caused by mechanical damages, removals, loss of materials depending on decay. Lost material (in particular ashlars and blocks) is often to be found near the building, maybe reused as building material. This kind of problem concerns mainly abandoned buildings.

- *Lichen, Moss* and *Plant*—depending generally on humidity or water leaks, they derive from morphological features of the surface, whose roughness favours the deposit of microorganisms, organic material and earth. They can cause the disaggregation of the stone: lichen's rhizines and plant roots may penetrate deep respectively into stone (ten to several millimetres) or into joints and fractures, weakening

data sheet D05

DISCOLOURATION - MOIST AREA

family

crack & deformation

detachment

features induced by material loss

discolouration & deposit

biological colonization

definition

Change of the stone coulor in one to three of the coulor parameters: hue, value and chroma. […]

Moist area: corresponds to the darkening (lower hue) of a surface due to dampness. […]

most likely reasons & annotations

Rising damp; pipe leakage; localized water leaks.

The discolouration of the surface due to humidity is only an aesthetical problem as long as soluble salts melt in the water don't crystallize; however, permanent dampness can lead to biological colonization.

material data connections

M01 local tuff stone	M05 concrete
M02 lime plaster & lime mortar	M06 metallic elements
M03 cement plaster & cement mortar	**M07 wooden elements**
M04 coatings	M08 glass & ceramic elements

technical sheets connections

TS01A paving - restoration	**TS06C cleaning - microorganisms**
TS01B paving - creation	TS07 replacing of a disaggregated plaster
TS02 collection of garbage	TS08 repositioning of missing stone elements
TS03 restoration of a covering system	TS09 repointing of joints
TS04 creation of a rainwater collect system	TS10 restoration of an iron element
TS05 reorganisation of the electrical system	TS11 restoration of a historic iron element
TS06A cleaning - incoherent deposit	**TS12 check of a wooden structural element**
TS06B cleaning - efflorescences	TS13 restoration of a wooden door - window

Fig. 3.18 Data sheet D05

data sheet D08   EFFLORESCENCE/SUBFLORESCENCE

family

crack & deformation

detachment

features induced by material loss

discolouration & deposit

biological colonization

definition

Efflorescence: Generally whitish, powdery or whisker - like crystals on the surface. Efflorescences are generally poorly cohesive and commonly made of soluble salt crystal.
Subflorescence: Poorly adhesive soluble salts, commonly white, located under the stone surface.

most likely reasons & annotations

Evaporation of saline water present in the porous structure of the stone and consequent crystallization of soluble salt on the surface or under the surface (depending on the speed of crystallization).
Former applications of cement mortar have amplified the presence of soluble salt and this kind of decay phenomena.

material data connections

M01 local tuff stone	M05 concrete
M02 lime plaster & lime mortar	M06 metallic elements
M03 cement plaster & cement mortar	M07 wooden elements
M04 coatings	M08 glass & ceramic elements

technical sheets connections

TS01A paving - restoration	TS06C cleaning - microorganisms
TS01B paving - creation	**TS07 replacing of a disaggregated plaster**
TS02 collection of garbage	TS08 repositioning of missing stone elements
TS03 restoration of a covering system	TS09 repointing of joints
TS04 creation of a rainwater collect system	TS10 restoration of an iron element
TS05 reorganisation of the electrical system	TS11 restoration of a historic iron element
TS06A cleaning - incoherent deposit	TS12 check of a wooden structural element
TS06B cleaning - efflorescences	TS13 restoration of a wooden door - window

Fig. 3.19 Data sheet D08

materials. Vegetal living beings grow where water is accessible, that is almost everywhere on exposed elements.

- *Soiling*—usually particles transported by running water, or heating convection, having different degrees of cohesion and adhesion to the substrate and depending on the roughness of plaster and superficial layers. This problem is common for the most part of surfaces.

Material and decay technical sheets are strictly connected with best practice sheets in order to have a global view of the analysis. At the same time, this permits to relate materials with pertinent decay phenomena and possible solutions.

Some details about the best practice interventions will be described in the next Chap. 4.

References

1. ICOMOS (2008) The Icomos charter for the interpretation and presentation of cultural heritage sites. http://www.international.icomos.org/charters/interpretation_e.pdf
2. Augelli F, Khachatourian Saradehi A, Khachatourian Saradehi L (2015) Anipemza: from genocide orphans' village to workers village. First proposals for conservation, valorisation and improvement of an interesting architectural settlement example and of a rich history site in Armenia. Scientific Papers of NUACA. IV (59):14–28, Yerevan
3. Stylianidis E, Remondino F (2016) 3D recording, documentation and management of cultural heritage. Whittles Publishing, Scotland, UK
4. ICOMOS (1996) Principles for the recording of Monuments, groups of buildings and sites. https://www.icomos.org/charters/archives-e.pdf
5. ICOMOS (2003) Charter on principles for analysis, conservation and structural restoration of architectural heritage. https://www.icomos.org/charters/structures_e.pdf
6. ICOMOS (2008) Illustrated glossary on stone deterioration patterns. http://www.international. icomos.org/publications/monuments_and_sites/15/pdf/Monuments_and_Sites_15_ISCS_Glossary_Stone.pdf

Chapter 4
Levels of Intervention, Conservation and Operational Guidelines for Anipemza Village

Paola Bertò and Alessandro Marcone

Abstract This chapter is focused on the definition of a conservation-oriented Operational Guidelines that help the local inhabitants and institutions in the urban and architectural management and to govern future transformations and maintenance operations, constituting a concrete reference for a feasible conservation of the qualities, the values and, indirectly, the rich memories of the Anipemza Village. These guidelines, inspired to the Principles of Restoration and to International Standards, highlight the need to consider three levels of intervention: the village, the buildings and the materials. At the end of the chapter some examples of the Operation Guidelines' technical sheets are presented.

Keywords Conservation · Guidelines · Recommendation · Urban preservation · Levels of intervention · Best practices

4.1 Introduction

The actual situation of Anipemza has been already described as unsatisfactory: uncertain housing conditions, precarious state of preservation of buildings and neglected common areas combine with unemployment and decline of population. Travelers, who annually come in their thousands to visit the Yererouyk Basilica's ruins, could be a significant driving force for the rehabilitation of the town, but today, a part from the absence of tourist facilities and the carelessness of the place, they would not find any information about the history of the village, about its population or about the activities here performed.

It would be naïve to imagine Anipemza to be a tourist attraction without making the locals aware of it; it is important to provide the formation and the education of the inhabitants about the development process, which should include the enhancement of the quality of the town, of the living conditions and of the working situation.

Preserving and rehabilitating of historical buildings is a good way to re-discover the local identity: the consciousness of their roots and the knowledge of their complex history, witnessed also by the architecture of the village, will strengthen the citizens' sense of belonging, stimulating at the same time the social and economic development [1]. To this aim the presence of the quarry could help: the tuff could be used in

© The Author(s), under exclusive license to Springer Nature Switzerland AG 2021
F. Augelli et al., *Preservation and Reuse Design for Fragile Territories'
Settlements*, SpringerBriefs in Applied Sciences and Technology,
https://doi.org/10.1007/978-3-030-45497-5_4

restoration works, lime should be produced on site... The revival of the traditional techniques and the training of specialized personnel among the inhabitants could also strengthening the local economy.

A municipal regulation on urban intervention would be desirable: the absence of rules permits inhabitants to act freely on buildings without any plan, adapting them to improve their living conditions. At the same time common areas are abandoned or used for private purposes. As a consequence, spontaneous interventions, additions, unattended debris and lack of maintenance are progressively changing the village's appearance and authenticity [2] (Fig. 4.1).

Fig. 4.1 View of a spontaneous intervention: an addition modifies the original volume of the building. The satisfaction of needs (new toilets, new windows and doors...) must not be in contrast with the conservation of the authenticity of the village

4.2 The Guidelines

Because of the economic difficulties and fragility of this territory, the aim of this work is to provide the local inhabitants with suggestions for low-cost though respectful way to operate on existing buildings.

Moreover, these guidelines should be addressed to different stakeholders at different levels: Regional offices, Municipality, Citizens/owners so that any intervention best-practice suggested is comprehended and shared through the different decisional levels. The guidelines take inspiration from similar works, whose pivotal phase is the attempt to cooperate with inhabitants in preserving local residential architecture, considered as testimony of their roots. Goal that can be reached through the education of local authorities, technicians and inhabitants.

An accurate knowledge of the historical built environment is the basis of the work, as already written in the previous chapter: not only the single building (or the single material) is analysed, but also—first of all—the connections between buildings which create the urban space. The guidelines act therefore on three different levels, from the urban scale to the single materials [3].

4.2.1 Level 1—Historical Village

Original connections among buildings create the distinctive nature of a place: their organic unity shall be safeguarded. Every building shall be seen as irreplaceable and no variations of the spatial relationships shall be carried out. Urban elements characterise the morphology of the town: some of them were designed, others belong to the local memory, so they shall not be modified. Common parts and private parts both create the urban form and must be maintained. An evaluation must be conducted by the municipality when an urban element has to be modified or removed, because of its deterioration: the new object shall match and respect the existent (Fig. 4.2).

Fig. 4.2 A view from the main street pedestrian walkway. Even if threatened by the absence of any maintenance guidelines the village is still rich of visible interesting details as long as an important memory to be preserved

4.2.2   Level 2—Historical Buildings

Historical buildings are characterised from the originality of material, shape, volume, inner and outer spaces; no alteration of the original features shall be carried out. The conservation is facilitated by use: however, the assigned function needs to be compatible with the structure, the inner distribution and the materiality. It is really important to understand that the authenticity of the village depends mainly on the authenticity of buildings. Inhabitants shall match the satisfaction of their needs with the protection and the conservation of their property.

New volumes cannot be inserted on common areas: only open elements with social aims (meetings, grill…) are permitted, and at the same time rusty and deteriorated ones must be removed.

New additions on private parts must be controlled and approved by the municipality: however, they must not be higher than the existing ones and have simple and linear volume.

Traditional materials must be preferred for preservation interventions for compatibility purposes but considering new additions these materials could be used as well, but using a more contemporary language for recognizability aims (Fig. 4.3).

Fig. 4.3 Wrong materials and poor choices during buildings' maintenance or upgrades might lead to substantial architectural quality loss: in this case wooden windows were replaced with PVC windows, featuring smoked glasses, causing unnecessary and improper alterations to the façade's details

Fig. 4.4 Original tuff stone and lime plaster were often covered/substituted with new materials, such as coatings and cement plaster. This is an example of a negative intervention because of compatibility issues

4.2.3 Level 3—Historical Materials

Materials are the only physic place where the Story sediments and leaves trace of its passing. The stratification incorporates different construction phases and every layer witnesses a determinate period: every modification shall thus be avoided. Removals are only permitted either when a material cannot be restored, because of its deterioration, or when it represents a danger to the preservation (Fig. 4.4).

4.3 General Standards in Conservation Works

International standards can inspire correct and acknowledged methodologies for the preservation of historical buildings. It is important in this sense to explain the basis of correct conservative interventions to the inhabitants and sensitize anyone approaching an intervention to the village's architectures about the best practices to follow. International standards are also available on the ICOMOS, ICCROM and Getty Research Institute home page, but every preservation program has to be strictly related with the real context of the involved place. However, in every situation a regular and programmed maintenance is essential to reduce the speed of the unavoidable decay effects.

From the aforementioned standards the following general guidelines can be learned and applied to the village's context:

- The use of compatible materials should be favoured.
- *Minimum intervention* logics should inspire every operation.
- Reversible solutions should be preferred.
- Every addition should be recognized as a contemporary expression.
- Do not remove original and historic elements and materials, if they can be preserved
- Do not replace original materials with modern ones: replace the missing parts with compatible materials, which shall have similar colour and texture.
- If the replacement of an element is technically needed, the new one shall both adapt to the context and be identifiable as modern.
- Do not assign to an historic building an incompatible function. If a new function is not compatible, choose another existing building or change function.
- Do not modify the façade of a building with the alteration of the shape of doors and windows.
- Check the structural conditions of the building regularly and consult a specialist when needed.
- Do not demolish original volumes or part of original volumes.
- Do not fence common spaces to grow private gardens: an area will be appositely designated.
- The use of second-hand elements is allowed, but poor building material (rusty metal, asbestos cement, slabs…) must be forbidden.
- Document any work through reports and photographs.

4.4   Anipemza's Guidelines and Technical Sheets

Guidelines for Anipemza Village have been organised as follows: the problem is analysed and described in the first sheet, where photos are provided, to explain what is adequate and what is not. The successive technical sheet illustrates step by step how the municipality or the owner has to act to solve the problem or at least improve the conditions. If a technical guide is not indispensable, some rules to be applied are given. When the problem cannot be easily identified or solved, the presence and the advice of a technician is strictly recommended. Every technical sheet is connected to material and decay sheets, to better show the relations between element, pathology and intervention. It also suggests the security devices to be mandatory used by workers. Observed relevant difficulties in Anipemza are described below, connected with the related technical sheets. The analysis starts from urban problematics (roads, garbage collection, covering systems, water collection systems…) and reaches the single elements and materials of private buildings. In closing some general indications about the care of the interiors are given. Below, Technical Sheet 01A, TS01B and TS03 are shown to explain the working and analysis method. The framework of the text does not permit the insertion of the guidelines in their entirety, so the remaining problems are listed and shortly described further (Figs. 4.5, 4.6, 4.7, 4.8 and 4.9).

UNEVEN AND/OR MISSING PAVING

Do not let the original stone slabs of the roads be lost or buried in the ground; if the historic paving is uneven and needs to be restored, use the local stone in the rehabilitation. Reuse, if possible, the existing stone slabs and produce handcrafted the missing ones.

Do not let driveways unpaved: cars and heavy vehicles damage the ground, causing holes that, with the rain, become puddles, amplifying the discomfort. Create a new paving using local stone.

Do not replace the original pavement or create new roads using asphalt, which is not a traditional material and modifies the breathability of the ground, amplifying humidity phenomena. If a new path is needed, create it with the use of local stone slabs. If a tarmac road needs to be restored, remove asphalt and replace it with local stone slabs. Only original asphalt from the Sovietic Period must be maintained and, if necessary, restored.

Fig. 4.5 Identification of the problem: uneven and/or missing paving

TS01A PAVING - RESTORATION

the problem

The paving is damaged by the passage of modern means of transportation: the stone slabs sink in the ground, which presents subsidence phenomena caused by the traffic and amplified by weathering. The creation of a new reinforced subgrade is suggested, but a regular maintenance can also be useful in ensuring the good conditions of the roads. If heavy traffic damages the stone pavement of the main street, an alternative route for such traffic must be considered; only perimetral roads will be used to this aim.

the intervention

A. Localise the damaged areas, remove the superficial deposit and the deteriorated slabs; remove the old bedding to reach a compact subgrade; the area of replacement must be larger than the damaged one, to facilitate the conjunction between the new and the old paving.

B. Recreate the bedding: use washed crush rock sand (ø 3 - 6 mm) to create a 4 – 5 cm thick layer.

C. Set down the slabs, following the original pattern: use the old slabs, if they are not too damaged.

D. Clean the surface using scrubbers.

E. Fill the joints with sand (ø 3 - 6 mm), spreading it on the surface with scrubbers.

F. Beat the paving using an iron pestle or a vibratory plate to ensure a good uniformity of the slabs. Water the surface and beat further.

G. Seal the joints with fine sand (ø 0 - 4 mm): spread the sand on the surface using scrubbers, ensuring the complete filling of joints. After 15 days, remove the surplus sand.

care of

municipality   owner

material data connections

M01 local tuff stone
M02 lime plaster & lime mortar
M03 cement plaster & cement mortar
M04 coatings
M05 concrete
M06 metallic elements
M07 wooden elements
M08 glass & ceramic elements

decay data connections

D01 blistering	D09 film
D02 fracture	D10 graffiti
D03 craquele	D11 lichen
D04 deposit	**D12 missing part**
D05 moist area	D13 gap
D06 staining	D14 moss
D07 disintegration	D15 plant
D08 efflorescence	D16 soiling

security devices

TETANUS

Fig. 4.6 Technical sheet 01A with indications and suggestions

TS01B PAVING - CREATION

the problem

Unpaved streets create discomfort: the creation of a new paving is necessary to avoid the problems caused by weathering (puddles, dirty roads…). Asphalt must be forbidden and the use of local stone strongly recommended, in the common parts as well as in the private parts. Self-locking stone blocks can also be utilized, but appropriate ones must be chosen.

the intervention

A. Set down the curb: the elements will be partially inserted in the ground, to delimitate and design the new road.

B. Create a compact subgrade (roadbed), with the use of coarse aggregate on the ground and medium aggregate on the top (15 cm thick layer), and the bedding, with the use of washed crush rock sand (ø 3 - 6 mm) to create a 4 – 5 cm thick layer.

C. Set down the slabs.

D. Clean the surface using scrubbers.

E. Fill the joints with sand (ø 3 - 6 mm), spreading it on the surface with scrubbers.

F. Beat the paving using an iron pestle or a vibratory plate to ensure a good uniformity of the slabs. Water the surface and beat further.

G. Seal the joints with fine sand (ø 0 - 4 mm): spread the sand on the surface using scrubbers, ensuring the complete filling of joints. After 15 days, remove the surplus sand.

care of

municipality   owner

material data connections

M01 local tuff stone
M02 lime plaster & lime mortar
M03 cement plaster & cement mortar
M04 coatings
M05 concrete
M06 metallic elements
M07 wooden elements
M08 glass & ceramic elements

decay data connections

D01 blistering	D09 film
D02 fracture	D10 graffiti
D03 craquele	D11 lichen
D04 deposit	**D12 missing part**
D05 moist area	D13 gap
D06 staining	D14 moss
D07 disintegration	D15 plant
D08 efflorescence	D16 soiling

security devices

Fig. 4.7 Technical sheet 01B with indications and suggestions

DAMAGED OR ABSENT ROOFING

Damaged roofs must be immediately repaired. Ruined or missing elements (gutter, pipes, stone and wooden cornices...) must be restored to avoid water dripping on walls, and consequently staining and biological colonization. If a tin roof is strongly eroded, leaky or dangerous, replace it with a new one using wooden structural elements and galvanised iron sandwich panels. The insulation layer (foam, fiberglass, cellulose fiber or plywood insulation) significantly increases the performance of the roof.

Do not leave inhabited or reusable or buildings without a proper covering: the absence of the roof causes a strong acceleration of the degradation.
When installing the new covering, remove only corrugated panels: original asbestos-cement rhomboidal elements must be encapsulated in the new roof.
Be careful when getting on the top of the roof, because a damaged roof could hide a ruined inner structure: check the roof structure from inside before!

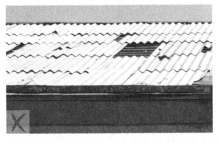

Tin roofs, when damaged, must be replaced with new galvanised iron roofs. Contact expert personnel only: the installation of the panels must be done in a workmanlike manner. Bad installed sandwich panels can deteriorate in a short time, due to weather and the frequent strong wind. Any lack of the covering can cause damage to the internal wooden structure. Waiting for the new galvanised iron roof, cover them with flashing.

Fig. 4.8 Identification of the problem: damaged or absent roofing

TS03 RESTORATION OF A COVERING SYSTEM

the problem

Lacks on roofs, damaged gutters and cornices lead to a general weakening of the covering, due to infiltrations of water. Ruined elements must be repaired or replaced. Disrupted chimneys are dangerous and must be rebuilt, to avoid elements falling on the street.
Wooden cornices must be restored.

the intervention

A. Locate the leaks and cover them with flashing, cutting it larger than the ruined area and fixing it with nails to the surface. This is only a temporary solution!

B. Clean and check gutters and pipes to ensure the correct drainage of meteoric water; when necessary, replace damaged gutters with new ones. A regular cleaning of gutters must be planned. Make sure pipes are correctly connected to the walls.

C. Replace damaged wooden cornices with new ones, connecting old and new cornices with iron plates. Apply an appropriate product on wood, to improve its resistance to atmospheric agents.

care of

municipality   **owner**

material data connections

M01 local tuff stone
M02 lime plaster & lime mortar
M03 cement plaster & cement mortar
M04 coatings
M05 concrete
M06 metallic elements
M07 wooden elements
M08 glass & ceramic elements

decay data connections

D01 blistering	D09 film
D02 fracture	D10 graffiti
D03 craquele	D11 lichen
D04 deposit	**D12 missing part**
D05 moist area	D13 gap
D06 staining	D14 moss
D07 disintegration	D15 plant
D08 efflorescence	D16 soiling

security devices

TETANUS

Fig. 4.9 Technical sheet 03 with indications and suggestions

4.5   Anipemza's Guidelines and Technical Sheets: The List

- *Uneven and/or missing paving*

    - *TS01A Paving—Restoration*
    - *TS01B Paving—Creation*

- *Garbage and unused building material*
  Do not deposit waste, unused building material, garbage in unsuitable areas or
  directly on the ground, to avoid the pollution of groundwater by leachate infiltration into the soil. Localize asbestos—cement elements, cover them when necessary, do not re-use them and await the removal by specialist. Do not enter rooms
  where asbestos—cement elements are disaggregating and pulverizing.

    - *TS02 Collection of garbage*

  Garbage and unused building material must be collected in designated areas.
  The use of a part of the unexploited quarry to this aim is strongly recommended. The indispensable number of rubbish bins, differentiated according to
  the garbage, must be arranged in the area and, adapted to the needs, in the village:
  a regular transport of the garbage to the main depot and, regularly, to the closer
  city/incinerator must be organized. Dwellers must be informed about it and have
  to respect the municipal rules in that regard.
- *Damaged or absent roofing*

    - *TS03 Restoration of a covering system.* See Fig. 4.4

- *Reorganising water and electrical system*
  Do not let meteoric water coming from gutters or waste waterflow along the
  perimeter of the building and infiltrate into the ground: that can amplify the
  problem of the rising damp and cause subsidence of the ground.
  If new electrical installations are needed, they must not interfere with the historic
  features of the building. Do not use metal junction box outside, because rusty
  boxes cause stains of walls and, when degraded, do no protect cables from meteoric
  water. Use a plastic one instead, and put it in the less visible place.

    - *TS04 Creation of a rainwater private collect system*

  Existing systems must be repaired and regularly maintained: if not large enough to
  accommodate the normal drainage, a new one can be made. Meteoric water must
  be used for agricultural purposes, organizing private rainwater collect systems
  using big tanks: they must be positioned in open spaces (not on roofs) or connected
  to the drainage system.

    - *TS05 Reorganising the electrical system*

  The electrical system shall be regularized to avoid cord dangling from the walls
  and the decay of the exposed cables due to weathering. External cables and wires

shall pass through waterproof cable ducts, similar in colour to the walls: their position shall follow the main geometric lines of the building. Internal wires must be gathered together with zip ties, or fixed on walls, avoiding them come in contact with water or pass through moist areas.

- ***Cleaning the surfaces***

  A regular cleaning of surfaces and elements of buildings is important to avoid the formation of hard deposit on materials: the absence of pollution in Anipemza makes the operation easier. Generally incoherent deposit, efflorescence and biological colonization can be removed using mechanical method; on the contrary, graffiti or stains due to rusty metal need chemical methods. An inappropriate cleaning can compromise a disaggregating plaster or mortar, so all surfaces must be checked beforehand: when the material is original, it must be consolidated before the cleaning.

  – *TS06A Incoherent deposit*

  Dust, soil and disaggregated material can be easily removed from surfaces, using water only when possible (absence of efflorescence) and vegetal brushes, whose hardness shall be chosen depending on the surface.

  – *TS06B Efflorescence*

  Efflorescence depends on the presence of water: a preliminary evaluation is mandatory to better understand the problem (rising damp, water leaks…). If the problem is not localized and not resolvable soon (e.g. rising damp from the ground) an ordinary maintenance can help in keeping the building in decent conditions. Surfaces must be brushed from top to bottom, removing crystalized salts. When needed, a new lime or microporous plaster can be applied: then new layer must match the level of the existing part, but simultaneously must be recognizable as new.

  – *TS06C Microorganisms*

  Microorganisms depend on humidity: when the cause of the presence of water is determined and eliminated, the application of biocides can generally solve the problem. Such preparations have to be compatible with the substrate, be degradable, not modify the colour of the surface, not leave residues or cause the formation of by-products, such as soluble salt. Surfaces must be brushed before and after the application of the biocide, to remove residue of organisms and product. Water must not be used in presence of efflorescence.

- ***Disaggregated/detached plaster***

  Plaster is a very important element of the building, because of its protective action. The protection is physical, because a missing or disaggregating plaster allows water to enter the joints, amplifying the decay; it is also chemical, because the basic PH performs a disinfection action. Disaggregated lime plaster must never be replaced with a cement one: this would amplify the decay conditions, because of the presence of soluble salt in the cement. If the crystallization of soluble salts carried by rising damp is the main problem two solution can be found: the creation of a ventilated hollow space along the foundation, which facilitate the evaporation of the terrain humidity or the application of a salt-resistant microporous plaster, to be replaced when damaged.

– TS07 Replacement of a disaggregated lime plaster

The disaggregating layer must be gently taken away, smoothing the edges of the original layer. Dust, debris, pulverized material must be removed from walls before applying the new lime plaster, which must be physically compatible with the old one and must have same colour and granulometry. To uniform the new parts with the old ones a natural colouring paint or lime milk shall be applicated.

- **Missing stone element**
Replace removed or fallen stone elements with the original ones, which are often nearby the building. If an original element is deteriorated or lost, recovered stone element shall be used. If the use of a new element is necessary, the new one must be physically compatible with the historic substrate and recognizable.

– TS08 Replacement of missing stone element

The absence of a stone element is generally caused by the disaggregation of the deeper mortar (due to the presence of water) and the subsequent detachment of the ashlars.
After the removal of the old mortar from the joints, the surface must be cleaned and wetted: the new element (shaped if needed) must be fixed with lime mortar and propped up with a wooden structure. After that, vertical and horizontal joints must be filled.

- **Incoherent stone walls' mortar joints**
The mortar is the weaker material of the walls and can easily disaggregate. Disaggregated mortar joints must be repointed, to avoid the amplification of the problem and the consequent weakening (even the collapse) of walls.

– TS09 Repointing of joints

The old mortar must be removed until reaching the sound part; sound mortar must not be removed. Cavities must be cleaned and wetted, before the filling of the joints. The mortar must be compressed to assure intimate contact with the blocks. To distinguish the new part, joints must not be squeezed or extruded, but rather flush or slightly recessed.

- **Maintenance**
Remove rusty and dangerous fence from the streets, and provide a proper maintenance of metallic fences, to avoid rusting. These elements (fences, gates…) must adapt to the local context: when painting them, use shades of grey.
Check the trees along the street regularly and, if needed, remove the dangerous one; if a line of tree is planned, replace the removed ones.
Do not paint building, parts of buildings or elements with a colour that deviates from the peculiar ones. The rose-orange-coloured aspect of the local stone is

a typical feature of the village, and must not be hidden. Do not apply modern materials on original ones and do not remove any layer, if it firmly adheres to the surface.

A regular maintenance can improve the condition of wooden elements and extend their life: be careful to identify the presence of water causing staining, because wet wood deteriorates easily.

– TS10 Restoration of an iron element

Protecting iron from rust means to protect walls and stone elements from fracture and permanent staining. After the removal of the old coating and of the rust with a paint remover and metallic brush, an anti-rust product must be applied on dried elements.

– TS11 Restoration of a historic/original iron element

Original elements have historic significance and must be preserved. A rust converter shall be used, to avoid the removal of the rust and the consequent loss of original material. Rust converter is expensive, so it has to be used only on relevant elements.

– TS12 Check of a wooden structural element

The condition of wooden structure must be checked regularly: a close observation and the use of simply manual instruments are recommended. The surface must be observed to monitor the presence of insects or humid areas: it must also be hit to verify the presence of inner holes. After a gentle cleaning a treatment must be applied.

Big fractures have to be sealed with a resin-based product: this non-rigid material allows expansions and contractions due to humidity.

– TS13 Restoration of exposed joinery

Alterations on wooden elements are normally not a problem, because incoherent deposit and stains can be removed or reduced through an accurate cleaning. Small fissures, instead, can favour the access of pathogen agents, so it is better to seal them. A protective coating must be applied (on outer elements a UV resistant one).

- **Interiors**

Provide a proper maintenance of internal wooden elements, not only doors and windows but also handrails and floorings. They do not need to be protected against weathering; therefore, minimal regular interventions and the presence of a varnish layer are generally sufficient to maintain them in good conditions.

Common parts of residential buildings must be kept clean: cleaner spaces are healthier and allow to observe early damage to elements (blistering or fissures on walls, cracks in wooden elements, wear and tear of floorings…).

Never use cement to fill the gaps caused by broken or absent tiles. Elements of different styles must not be mixed: existing/original floorings must be restored, if needed.

Inner spaces and proportions must not be modified through the demolition of original walls or the construction of new ones. If the modification of the size of an inner opening is necessary, a technician must be involved, to avoid structural problems.

Homes are normally heated by stoves, but ducts, useful to heat spaces, must not pass at eye level. Original walls can be pierced only if strictly indispensable.

References

1. Environmental Operation Unit (1999) Cultural Heritage Guidelines. A handbook for staff and contractors, Walkerville
2. ICOMOS (1994) The Nara document on authenticity. http://www.icomos.org/charters/nara-e.pdf
3. Gianbruno M, Pistidda S (ed) (2015) The walled city of Multan. Guidelines for maintenance, conservation and reuse work. Altralinea Edizioni, Firenze

Chapter 5
Anipemza Village's Adaptive Reuse Design

Francesco Augelli and Matteo Rigamonti

Abstract This Fifth Chapter presents the adopted methodology and the results of the Adaptive Reuse Design that, starting from the outline of a social-oriented System Design Masterplan and the priority list of suggested operations is dedicated to the presentation of the case-studies that compose the mosaic of those interventions needed to build a possible social and economic reactivation strategy of Anipemza. Five case-studies for five different existing buildings will in fact virtuously match the user needs of locals and visitors alike, following preservation priorities to pursue the conservation of the integrity and the authenticity of the pre-existences, a minimum intervention approach and contemporary interventions identity's awareness. The aimed attraction of a larger number of tourists into the discovery of the architectural, but most importantly the human qualities of the village will endorse decisive new social relationships as long as the village micro-economy reactivation, thus endorsing the future itself of this community, the historical memories and the architectural palimpsest. In this sense the Adaptive Reuse Design is intended as the final stage of the Conservation Process.

Keywords Adaptive reuse design · Strategic reuse masterplanning · Social reactivation · Micro-economic reactivation · Heritage innovation

5.1 Adaptive Reuse Strategies' Outline and Motivation

The Adaptive Reuse Design Project for Anipemza considers the reactivation of the village's buildings as the last stage of the Conservation process [1], suspended between the preservation of the memories and the architectures from one side and the insertion of a set of compatible new functions tailored around today's user needs, being them locals or visitors alike.

The main objective of the present work is to first provide strategies for the social reactivation of the village, and therefore to create the conditions for a cultural and architectural preservation intervention projected in the future and intimately intertwined with the community's prospect [2].

In this sense the present design is certainly aiming at a prompt improvement of the dire living conditions standards today experienced by the citizens, especially in terms of hygienic conditions, at the same time is purposing strategies to endorse a micro-economic reactivation of the village, that is one of the major causes of Anipemza's abandonment as well.

As it will be explained in the next paragraphs, one of the major opportunities from this point of view is offered by tourism, whose introduction in the village's system, today nearly absent, has to be well balanced not to spoil the village's identity. In this sense the purposed Adaptive Reuse Design will be based on the objective of creating a virtuous relationship between the locals and the visitors where actions taken on both sides will possibly generate positive repercussions on the other, considering a user-centred approach. This is why the understanding and intertwining of the problems of the villagers and the tourists has suggested the design opportunities directions that will be described in Sect. 5.3 and in the following ones.

Within this context, while the result of the researches on the village user's urgencies has been explained in Sect. 2.4 and following, the understanding of the tourism's situation and needs will be presented in the next paragraphs, underlining those touch-points that will possibly trigger a virtuous micro-economic and social exchange between the locals and the visitors, endorsing the material and immaterial values, evidences and identity of this extra-ordinary village and context.

While the economic problems of the residents are strongly connected to the lack of profitable activities and revenues, the tourists[1] still need basic facilities such as a bathroom and an information point. These needs could be satisfied while offering food and local products, as well as the possibility to discover the qualities of the village. In this way a connection between the needs of the visitors and the locals might gradually start a virtuous process able to endorse the economic conditions of the Anipemza's people and to start more demanding interventions like roads' repair, water availability, the repair and improvement of residential buildings, the repair and reuse of other buildings for cultural and tourism's purposes.

The definition of a strategical reuse masterplan has more than a mere formal architectural preservation as a goal, it's more about the conservation of the village's identity and life, about looking for concrete propositions that might ensure the reactivation of a micro-economy, preventing citizens from leaving and, at the same time, raising their awareness of the historical and cultural values of their home town (Fig. 5.1). This will be the social and economic base needed to ensure the future life of Anipemza: a basic but solid ground[2] on which to grow possible further development's scenarios (Fig. 5.2).

This work identifies on one hand an existing tourist flux as a reasonable, concrete, existing resource yet to be exploited. On the other hand, the reuse project needs to work on a system of selected buildings' remodelling, offering functional and experiential solutions to both visitors and villagers alike.

---

[1] Several hundred people visit Yererouyk's ruins along the route to Gyumri every year, as better explained in Sect. 5.2.
[2] See Chaps. 2, 3 and 4 for more details about the village's people's needs and state of the arts' conditions.

Fig. 5.1 Two drone views of Anipemza. From East, approaching the village from Yererouyk ruins. (Top) From North, approaching the village from the tuff quarry. (Bottom) In both photos: The natural gorge of the Akhurian River sets the boundary between Turkey and Armenia in this area, located just few meters distant from the settlement. Photos captured in May 2019. Courtesy of Marco Germi

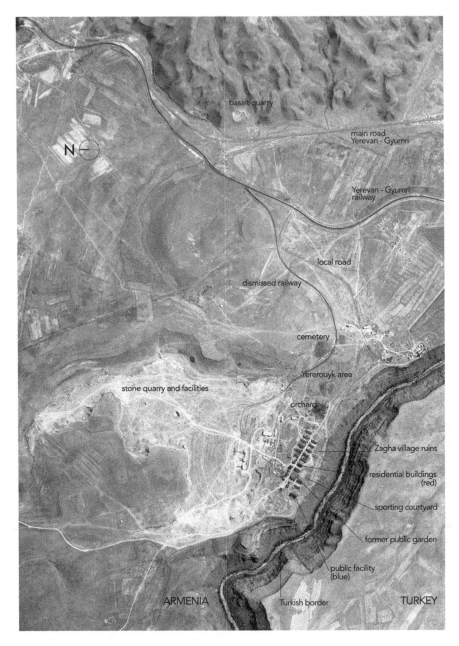

Fig. 5.2 The Anipemza Village as it appears from satellite (2015): territory, connections, morphology, facilities

The aim is to endorse the encounter between outsiders and the local community, considering tourism as a social opportunity that goes beyond the contemplation of a mere economic advantage to be taken. This is the reason why this research started from the analysis of tourism facts to draw a landscape in which Anipemza can be localized, trying to understand if the village may play a credible role into a larger scheme that assumes the existing Armenian tourists' flows as fuel for the whole reuse design's strategies. Consequently, we are to present the system of accurate interventions that defines our reuse masterplan.

The operations should in general consider the following different steps, set in order of urgency.

1. Creation of Guidelines for the village's maintenance to be distributed to the inhabitants: this material is aiming to explain why and how to avoid replacement in favour of repair and preservation to endorse the conservation of the existing trough accessible, low-budget, everyday interventions. The secondary but very important objective is to raise the local's awareness on the many qualities their village possesses and that any sensible alteration of the historical materials and typology might challenge the existing values and therefore the visitor's interests in Anipemza.

2. Starting micro-economy's reactivation oriented businesses, involving all inhabitants, offering organic and local quality products to tourists visiting Yererouyk ruins. These might include but not be limited to honey, bread and baked goods, typical sweets, lavash,[3] katha,[4] sujuk,[5] cakes, fruits (fresh and dried) and vegetables (fresh and in jar), jams, small carved stone objects. In addition, thematic lunches' experiences and guided tours to Anipemza and interesting surrounding sites including but not limited to archaeological sites, bird watching activities as well as trips to Kharkov, to look, even if from afar, to Ani, the ancient Armenian Capital might be offered to engage tourists. The possibility to provide accommodation for the night might be an interesting proposal too.

3. The establishment of a visitor centre to offer an information point, a bookshop, a coffee shop and bathrooms for the tourists in transit from Yerevan to Gyumri or asking for permits to visit Ani from Kharkov.

4. Communicate the achievement of the first three suggested steps to further enhance the awareness on the interesting and unique case of Anipemza to public opinion and national and international authorities.

5. Identification, during the first four steps, of national and international sources of funding for further interventions.

---

[3] Lavash is the typical Armenian bread that has been inserted in 2014 in the UNESCO World Heritage List.

[4] Katha is a typical cake made in different ways according to different family recipes, basically made of flour and eggs and a honey and nuts' stuffing.

[5] Sujuk is a dessert that is widely prepared in many Near-East Countries with some variations. It is a sausage-looking cake, basically made of nuts, grape must and flour.

6. Road's paving repair, improvement of water supply and power lines: Implant design for solar energy based self-sufficient systems, rainwater collection, artesian wells, geothermal system and other possible renewable energy system integration.

7. A Conservation Project Design, Consolidation and Compatible Integration of toilets, kitchens and all needed comforts in all buildings, to ensure contemporary living standards.

8. Reuse of abandoned buildings as Museum, Restaurant and Apartment for tourists and residents as well. Introduction of a diffused hotel system.

9. General reorganization of the public spaces and reconversion of the Quarry for tourism purposes and possible other revenue-generating uses. Connections with other activities, associations, sites in the area and along the Yerevan-Gyumri road might be very important.

10. Starting larger than micro-economic scale businesses in Anipemza, involving all the inhabitants, addressed to tourism as well as a community-run restaurant, residences for guests and a diffused hotel.

11. Creation of an international summer school for tuff sculpture tradition's preservation as well as promotion of the possible start-up activities to be endorsed on the territory, connected to intangible/cultural heritage preservation, principally related to stone-carving.

12. Improve minibus and taxi-shuttle routes for both tourists and residents.

13. Realization, inside unused buildings, of cultural activities such as: Museum and Study Centre of the extraction and processing of tuff in Armenia and Anipemza.

14. Photo exhibition of daily life in Anipemza in the past.

15. Museum of the Armenians' orphanage and orphans.

16. Museum of the labour camps and workers' villages in Armenia.

17. Museum of Yererouyk.

18. Museum of Ani.

19. Permanent exhibitions regarding Tamanyan and Toramanyan's[6] restorations and Yererouyk's studies.

As a general methodological approach to preservation it is important to underline that all solutions outlined in this chapter, as well as all Guidelines for future interventions, are based on the respect of the integrity of the buildings and all manufacts present in the village as well as their material culture. For this reason, as a general rule of action, addition has been always preferred to subtraction [3] to endorse the conservation, and possibly the communication, of all layers composing the rich palimpsest of the historical background of the village.

Furthermore, to ensure that authenticity will be preserved as a fundamental value, all purposed additions should be designed in formal and material harmony with the existing, as much limited to the necessary as possible, and identified as contemporary interventions, in accordance to the Principles of Conservation. See Chap. 4.

---

[6]See Chap. 2.

5.2 Tourism's Facts

5.2.1 Introduction to Tourism's Facts Analysis

The project for the conservation and reuse of Anipemza village started from an in-depth analysis of the current situation lived by the village's inhabitants, the more accurate as possible, even from the economic point of view. One of the assumptions of this research resides in the consideration that a plausible economic scenario should support the design proposal. For this reason, it is important to evaluate that Yererouyk archaeological site's proximity offers interesting future eventualities of development, as well as concrete opportunities that should be taken.

As a matter of fact, current tourists' fluxes prove that a new life for the village of Anipemza is possible, just because the reactivation might rely on existing, not theoretical, data.

Moreover, we think that two other factors are positively favourable.

First, the increasing notoriety of Yererouyk archaeological site, that in the present days witnesses a renovated interest by the international scientific community,[7] might help attracting investments in the near future and lead to a growing number of visitors.

On the other hand, we believe that the existing important, even if perfectible, infrastructures that nowadays connect[8] the capital city Yerevan with Gyumri, today the second Armenian city in importance and number of citizens, support the feasibility of the purposed masterplan. The key to reactivate a micro-economy that will permit Anipemza's inhabitants not to leave their homes for a better future resides in beginning little by little, but with a wider and more ambitious scheme in mind, a positive direction to take able to materialize a believable future for the village's conservation and its existence as a living urban fabric.

Motivated by this spirit we have run the most complete research we could on institutional websites, private and public statistical authorities and international tour operators to frame our reuse masterplan into existing tourism's market and fluxes.

5.2.2 Armenian Tourism: A General Overview

The project for the conservation and reuse of Anipemza village is based on the analysis of the situation as lived by the village inhabitants in the years 2015/16, and on the consideration that Armenian Tourism is a fast-growing business. This market

---

[7] Yererouyk Basilica Ruins first entered the UNESCO tentative list in 1995 as one of the earliest surviving Christian Monuments. On 16/03/2015 Yererouyk Archaeological Site and Anipemza Village have been inserted in the definitive list of the "7 Most Endangered European Sites" from "Europa Nostra ONLUS" and "EIB: European Investment Bank".

[8] Maintenance of the last 4 km long segment of the road leading into the village appears to be particularly urgent. This road's very bad condition is today the main reason why Yererouyk archaelological site is excluded from the majority of touristic tours' plans.

at the moment contributes to Gross Domestic Product for 530 million Euros, that is 4,7%, offering approximately 120.000 jobs [4].

These facts describe a still developing industry, especially outside the capital.

The situation appears evident to anyone traveling across the country, from the accommodation offer as well as cultural, landscape and architectural endorsement: only few of the many points of interest reveal an appropriate communication or museographic project, and almost no one can offer services that can be considered more than basic, or able to engage the tourists (Fig. 5.3).

Nonetheless, considering the last few years' data, it is evident that the sector is fast growing, with bright prospects for the future, even considering the difficult international political relationship with Azerbaijan regarding the Nagorno-Karabakh region and with Turkey.

Analysing the data regarding tourists' origins we can understand that the majority of them comes from the Commonwealth of Independent States, Russia and Georgia in particular (65%). Departures from European Community are second (20%), followed by Iran (10%) and United States of America (5%) [5].

Regarding travel typologies, pilgrimages connected to genocide and diaspora are preferred, representing the 65% of the whole tourist accesses of the country [5]: usually these travel routes include also religious places of interest on the territory, favouring Geghard, Echmiadzin e Zvartnots, Haghpat e Sanahin UNESCO sites.

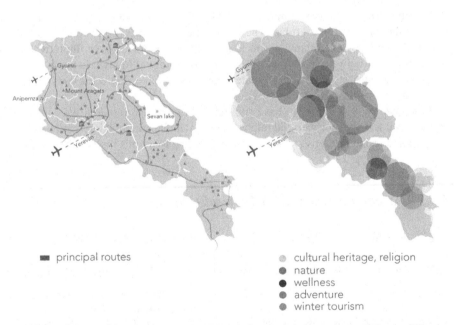

■■ principal routes

● cultural heritage, religion
● nature
● wellness
● adventure
● winter tourism

Fig. 5.3 From left: the general map of the main existing touristic routes in Armenia, the different touristic offer's typologies and their general distribution in the country

Tourism concentrated on wellness and/or naturalists' interests can be recently taken into account, especially in lake Sevan and Mount Aragats' areas; an ascending trend of trekking and off-road requests is also reported.[9]

In Mount Aragats' surroundings and in Jermuk's area, South-East of Yerevan, winter sports and skiing facilities are concentrated, these summon not only local tourist but also foreigners, especially from the bordering countries.

5.2.3 Existing Major Travel Routes and Anipemza

Cross-checking data from more than forty local and international tour operators we mapped the principal tourist routes in the region. Our research has been focused on testing how a possible stop in the Anipemza village could be inserted into existing tour schedules' proposals from different travel agencies.

As previously stated, Anipemza is located two kilometres far from an important highway connecting Gyumri and Yerevan. At the moment the two principal traveling routes are set[10] along this infrastructure, but there are sensible differences between the two, so that the situation needs to be more in-depth evaluated.

The first route consists of tourists coming from Yerevan and leading to Gyumri.

Following this direction, the travel time between Yerevan and Anipemza is more than two hours.[11] The majority of the tours leaves the capital early in the morning and plans to visit Echmiadzin and Zvartznots' UNESCO sites before passing by Anipemza, leading to Gyumri, around the very first afternoon. This could support the idea of a late lunch stop in Anipemza once a restaurant and/or bar facility will be set[12] (Fig. 5.4).

It is true though that there is plenty of other locations of interest[13] between the capital and Anipemza. A visit to one or more of these sites, once a Yererouyk stop will be possibly inserted into usual travel schedules, could delay the arrival in Anipemza area in the mid-afternoon, and therefore we can reasonably think that a dinner or even a night accommodation might be supposed once the strategies of this study will be realized, offering cultural facilities and attractions to tourists, possibly endorsing at the same time Yererouyk site importance.[14] The second consists of tourists coming from Gyumri and leading to Yerevan. The travel time between Gyumri and Anipemza

---

[9]A list of the considered tour operators can be found in the web references.

[10]The entirety of the organized Armenia's tours can be basically assimilated to a round trip route that can be travelled clockwise or counter-clockwise according to the details described further in the chapter.

[11]More or less two hours and thirty minutes, a sensible reduction of at least 15 min of this travel time can be considered once a major maintenance on the severely damaged two kilometers connection between Yererouyk ruins and the highway will be done.

[12]See Chap. 4.3 General Reuse Strategical Masterplan's definition.

[13]Confront Figs. 5.3 and 5.4 for more details on the possible endorsement of the area's points of interest.

[14]See Sect. 5.10 for possible future outcomes.

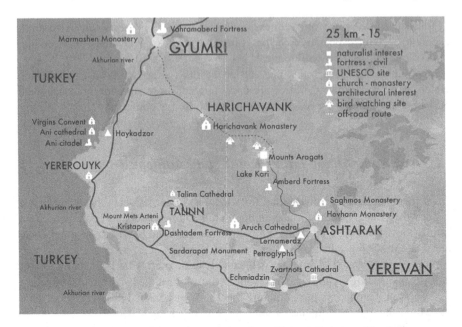

Fig. 5.4 Existing travel routes and main touristic points of interest in the considered area

is less than two hours.[15] In this case the arrival time depends largely from variables as visitor interests and the fact they have spent the night in Gyumri or not. We can assume, considering the tourist places of interest around and before Gyumri part of the counter-clockwise travel route, that a stop in Anipemza might occur at noon as well as in the late afternoon, supporting again the hypothesis stated in the previous case.

In conclusion, even considering that the harsh condition of the two kilometres road connecting Yererouyk site and the previously described highway,[16] that prevents for safety reasons the majority of the travel agencies and bus drivers from stopping in the area, Yererouyk site counts an average of two hundred visitors per week during the high season.[17] This fact means that the reactivation of a micro-economy is far from being an unattainable hope and this masterplan relies on an existing travel route, already serving the area. We believe that two kilometres road's maintenance and the creation of basic tourist facilities will attract even more tourist to endorse Yererouyk archaeological site's tourist flux, at the same time encouraging people to discover

---

[15]More or less one hour and thirty minutes, as already stated, a sensible reduction of at least 15 min of this travel time can be considered once a major maintenance on the severely damaged two kilometers connection between Yererouyk ruins and the highway will be done.

[16]Bus drivers nowadays discourage tourists and tour operators from reaching the Archaeological Site, due to the dangerous roads' conditions.

[17]These data come from a direct interview with Harutyun Tarlanyan, Anipemza village's Mayor in charge at the time of the survey. See also [6].

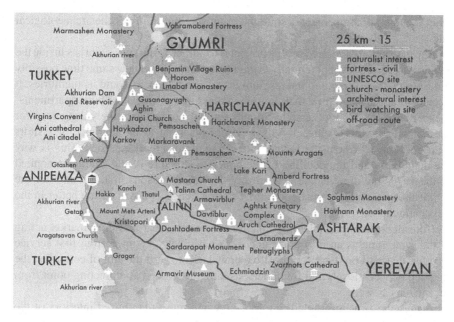

Fig. 5.5 Possible future touristic scenario in the considered area

the cultural and historical qualities of Anipemza. This might hopefully be the key to a feasible preservation of the village and a better future for its inhabitants (Fig. 5.5).

5.3 Strategical Reuse Masterplan's Definition

The objective of the Reuse Strategies presented in this paragraph is to define a Masterplan able to organize the complexity of the different aspects involved in the Adaptive Reuse operation, somehow exploiting its multi-layered fields of action.

Not only the reuse of historical materials, the continuation of cultural phenomena through built infrastructure, connections across the fabric of time and space and preservation of memory [7] are relevant aspects of this study. The adaptive reuse design masterplan of the village aims to implement the socio-economic strategies through the suggestion of simple but relevant contemporary experiences aiming the reactivation of the village, considering Anipemza's past and present richness as living matter, and connecting the different scales and intertwined aims of the purposed conservation and adaptive reuse intervention.

In the previous paragraph an existing tourist flow has been outlined. On a weekly base, visitors reach Yererouyk Basilica's ruins, but the only sporadic, casual contact occurring between the tourists and the locals does not ensure at the moment any valuable social exchange. Moreover, no revenue is gained since no service is offered from the town of Anipemza, despite the promising proximity of the archaeological site.

The first action suggested by this project is to attract Yererouyk's tourists inside the village inserting a set of potentially engaging new functions at the same time possibly arousing their curiosity and interest into discovering Anipemza and its memories. It is evident at the same time that the ruined condition of the latest built apartments' facade,[18] directly facing the archaeological site, is not just extremely dangerous for the inhabitants, but is currently also a major deterrent to the entrance of visitors as it does not represent the town's interesting architectural qualities. This is why the demolition and remodelling of the concrete structure of the balconies applied to the 1970s building facade is a priority: for safety reasons first, but also for the removal of a tangible obstacle to tourist's attraction, as we will define better in the next paragraph.

Considering the possible new functions to insert, a bar and mini-market acting at the same time as an info-point may be a basic yet very interesting first service to offer, that will produce very positive drawbacks toward the aim of endorsing the meeting between locals and foreigners, as it might be used as a meeting point from the citizens as well.

A museum is a fitting facility to preserve and reuse the unused interiors of the House of Culture that, through minor interior transformative interventions, will host not just valuable information about Yererouyk, the village and its history and other related topics, but may also host events that will animate the locals' everyday life (Figs. 5.6, 5.7 and 5.8).

A widespread hotel's rooms set in between the inhabitants apartments all around the village can become a service that may be run from the locals and become a new activity that might involve more than a few employees and become an engaging experience for the visitors who will stay for the night as well. At the same time this proposal will also enable to suggest how to reuse the buildings with low-impact renewal solutions for the inhabitants' interiors, that are nowadays in need of knowing how to deal with obsolete soviet interior distributions and plant upgrades of their possibly charming, but not fitting nowadays' needs, stone buildings.

The generally harsh sanitary conditions, mainly due to plant absence or deterioration in the buildings, requires to identify an area in the village where to set temporary sanitary facilities able to help the local community until the restart of an economy will take stage and will permit major plant maintenance on most of the buildings.

Lastly, the proposal of a restaurant inside the former Workers' Dining Hall may attract even more tourists into the village and offer a new life to one of the most ruined yet interesting buildings of Anipemza.

In the next paragraphs specific details about the Adaptive Reuse project operations will be presented following an order based on their position within the village: from the closer to Anipemza entrance and Yererouyk ruins to the farthest (Fig. 5.9).

---

[18]See Chap. 2.

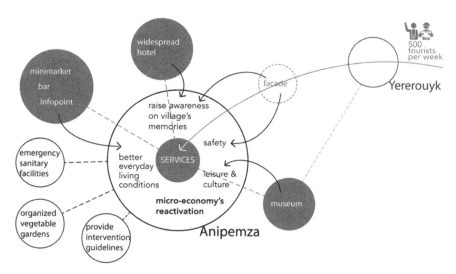

Fig. 5.6 General reuse strategic masterplan's diagram for the village's micro-economy reactivation. The dark grey bubbles represent the adaptive reuse's case studies described in detail in the next paragraphs. The arrows represent positive repercussions on the system. The new restaurant in the former dining hall is absent here representing a further step from the initial strategies

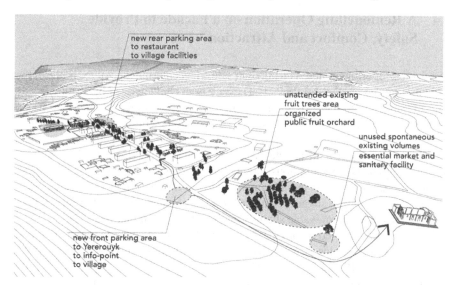

Fig. 5.7 Perspective sketch view of the general masterplan's strategies: new general village facilities' positioning

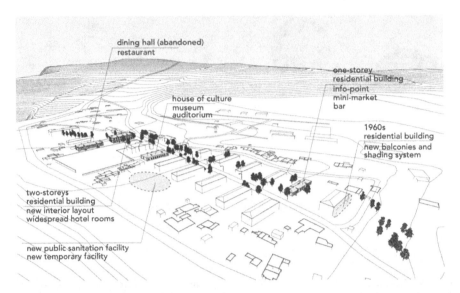

Fig. 5.8 Perspective sketch view of the specific masterplan's strategies: adaptive reuse's case studies' positioning

5.4 A Remodelling Operation on a Façade to Provide Safety, Comfort and Attraction

The first building encountered on the right side when entering the village is one of the two buildings erected in the 1960s.[19] They are both characterized by a stone facade but, differently from the second one, the first features a prefab concrete light frame structure connected to the front, realized to provide balconies and probably some direct sunlight's mitigation.[20] This structure manifests an evident state of decay and is dangerous not just for the people living inside the building. Its unstable conditions have been aggravated by additions: some of the balconies were closed to offer more space opportunities inside the apartments. This is the main motivation that led to the loggias' demolition suggestion and the design of a new freestanding structure to provide balconies and movable shutters to protect from summer heat.

The solution purposed is composed by the realisation of concrete foundations and an x-lam[21] solid wooden structure connected and reinforced by metal carpentry's solutions (Figs. 5.10 and 5.11).

The composition of the new façade has been inspired by the old structure's rhythm and positive and negative spaces' distribution. The exaggeration of some of the

---

[19]See Chap. 2 for further details.

[20]This assumption is based on the position of the building, facing South-East and with no shades or sight obstacles that might have induced the designer to provide a shading solution.

[21]X-lam is one of the massive cross-laminated engineered structure system's products made of wooden boards joined with formaldehyde-free adhesives.

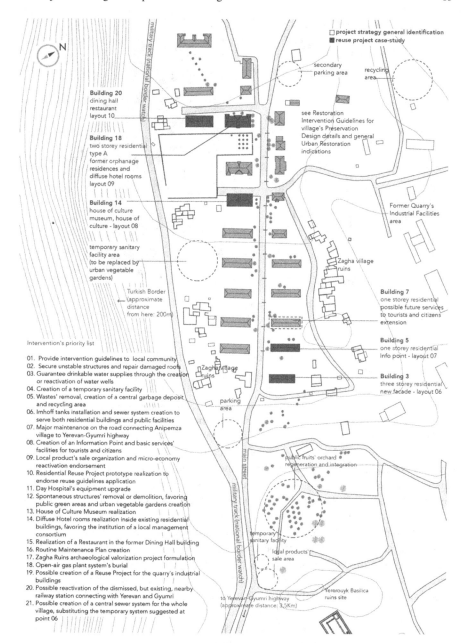

Building 20
dining hall
restaurant
layout 10

Building 18
two storey residential
type A
former orphanage
residences and
diffuse hotel rooms
layout 09

Building 14
house of culture
museum, house of
culture - layout 08

temporary sanitary
facility area
(to be replaced by
urban vegetable
gardens)

Turkish Border
(approximate
distance
from here: 200m)

Intervention's priority list

01. Provide intervention guidelines to local community
02. Secure unstable structures and repair damaged roofs
03. Guarantee drinkable water supplies through the creation
or reactivation of water wells
04. Creation of a temporary sanitary facility
05. Wastes' removal, creation of a central garbage deposit
and recycling area
06. Imhoff tanks installation and sewer system creation to
serve both residential buildings and public facilities
07. Major maintenance on the road connecting Anipemza
village to Yerevan-Gyumri highway
08. Creation of an Information Point and basic services'
facilities for tourists and citizens
09. Local product's sale organization and micro-economy
reactivation endorsement
10. Residential Reuse Project prototype realization to
endorse reuse guidelines application
11. Day Hospital's equipment upgrade
12. Spontaneous structures' removal or demolition, favoring
public green areas and urban vegetable gardens creation
13. House of Culture Museum realization
14. Diffuse Hotel rooms realization inside existing residential
buildings, favoring the institution of a local management
consortium
15. Realization of a Restaurant in the former Dining Hall building
16. Routine Maintenance Plan creation
17. Zagha Ruins archaeological valorization project formulation
18. Open-air gas plant system's burial
19. Possible creation of a Reuse Project for the quarry's industrial
buildings
20. Possible reactivation of the dismissed, but existing, nearby
railway station connecting with Yerevan and Gyumri
21. Possible creation of a central sewer system for the whole
village, substituting the temporary system suggested at
point 06

project strategy general identification
reuse project case-study

secondary
parking area

recycling
area

see Restoration
Intervention Guidelines for
village's Preservation
Design details and general
Urban Restoration
indications

Former Quarry's
Industrial Facilities
area

Zagha village
ruins

Building 7
one storey residential
possible future services
to tourists and citizens
extension

Building 5
one storey residential
Info point - layout 07

Building 3
three storey residential
new facade - layout 06

Zagha village
ruins

parking
area

main street

military track (national border watch)

public fruits' orchard
regeneration and integration

temporary
sanitary facility

local products'
sale area

to Yerevan-Gyumri highway
(approximate distance: 3,5Km)

Tererouyk Basilica
ruins site

Fig. 5.9 General reuse plan and complete intervention's priority list

Fig. 5.10 Building 3 in 2015. The loggias appear in a very bad condition representing a potential danger for the inhabitants

balconies' thickness is a dramatic effect that aims to underline and somehow preserve the memory of the positive and negative spaces' intervention of the spontaneous closings and offers the façade's shape more interesting shadows.

The wooden shutters' independent position and overlapping system offers customization's possibilities, as well as an aesthetical random vivacity (Figs. 5.12 and 5.13).

The choice of a fairly recent wooden structure system[22] helps distinguishing the original structure from the new layer of the contemporary addition. At the same time, wood was inspired by the many examples of traditional Armenian wooden loggias and balconies that can be seen in the village, for example on the façades of building 4,[23] or the buildings 6, 10 and 12 that share the same two-storey typology, or throughout the whole country by even larger, noteworthy examples.

This attractive landmark will be the statement that a contemporary intervention occurred in the village, clearly visible from anyone approaching and from the Yererouyk's ruins, and might be perceived by the visitors as an invitation to discovery.

Reinforced extra white glass balustrades may offer unexpected reflections and a vivid, contemporary-looking solution, but may be substituted with a simpler and budget-oriented wooden or metal railing system.

One of the two parking areas defined by the project is set just across the main street, but more downhill toward the archaeological area, to facilitate the access to

---

[22]The first massive wooden engineered systems, very different from wooden frames structures, were born in the 1990s.

[23]See Fig. 5.14 and Chap. 2 for more details.

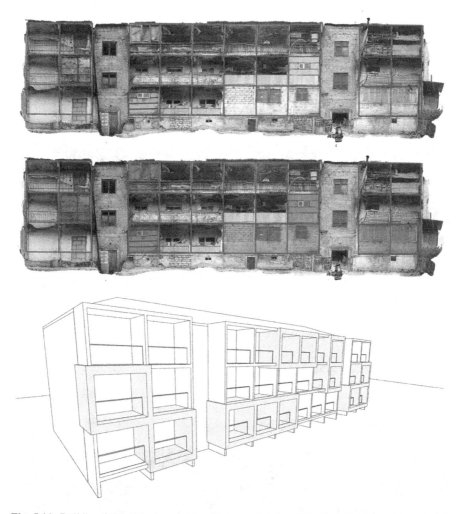

Fig. 5.11 Building 3 façade's remodelling. The new façade will replace the dangerous existing loggias. The new design is inspired by the building's proportions and considers the spontaneously added volumes occurred after its realization

both the ruin complex, the village entrance and the bar and info-point reception facilities that will be described in the next paragraph (Fig. 5.14).

5.5 The Info-Point Building: A Space of Encounter

A basic service structure to the tourists, today totally absent, offering a bar, sanitary facilities and some rest can be easily provided, and will be certainly appreciated by

Fig. 5.12 Building 3 façade's remodelling. View from Yererouyk's ruins in 2015 (top). Rendering view from Yererouyk's ruins simulating the project's intervention on the façade and the reorganization of the existing spontaneous fruit trees' area (bottom)

Fig. 5.13 Building 3 façade's remodelling. View from the village's entrance in 2015 (top). Rendering view from the village's entrance simulating the project intervention on the façade and on building 3 surroundings (bottom)

Fig. 5.14 Building 6 in 2015. View from the main street. The insertion of a wooden balcony enriches the façade's composition with a traditional Armenian architectural element that can be seen on many buildings and in different contexts through the whole Country

Fig. 5.15 Building 5 in 2015. View from the main street

any visitor, might be identified as the first functional intervention to be realized of the reuse's case-studies.

This project purposes that this facility may play an important role in many ways and suggests locating the bar not in the ruins' surroundings, which may be seen as a rational choice, but in the second building on the right encountered when entering the village.

Fig. 5.16 Spontaneous additions occurred in the whole village but particularly on one-storey buildings like the considered one

This position will ensure the entrance to the village, so that the basic services to visitors may be accessed only getting in touch with Anipemza's reality. On the other hand, this position will also see the bar as a plausible meeting point for the locals too, resulting in a gathering facility. The bar will become also a place where foreigners can actually be invited to get in touch with the local community, sharing a common space and being informed about all the activities and the services offered by the village. This is the reason why the project mixes the info-point facility's functions, that includes an office and a display area, with the bar: the intent is to cross the border between visitors and locals as much as possible.

Talking about functions and sharing common spaces, a mini-market could be inserted for all customers, allowing for a very comfortable store addressed to any considered users: souvenirs, local products' supply and everyday life's goods as well.

Fig. 5.17 Interiors' pictures shot in 2015 inside building 5 and 7

A restroom area is planned in the central part of the building.

The selected host structure is a one-storey former residential building that suits the spatial needs to collect all these functions together. Actually, its interiors are a sequence of rooms, each and every of them characterized by a proper soul and memories (Figs. 5.15, 5.16, 5.17, 5.18, 5.19 and 5.20).

This fact inspired all interior design's solutions that have been founded on the radical decision to preserve all the materials and the original characters of each room. New doors' opening will permit a fluid flow between the new functions resulting in a reverie travel experience through the memories of the residential singularities.

Fig. 5.18 Info-point building's adaptive reuse design. Perspective section of the bar area (top). Renderings: the market (middle), the bar and info-point area (bottom)

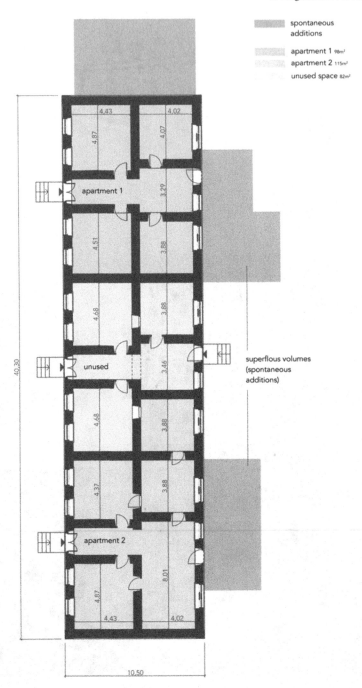

Fig. 5.19 Building 5 in 2017. Ground floor plan and existing functions' zoning

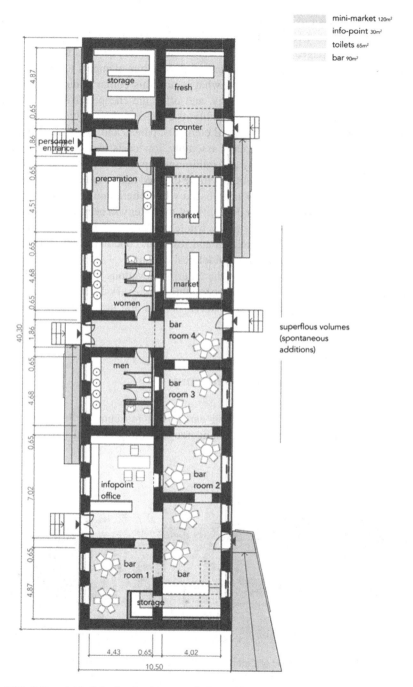

Fig. 5.20 Info-point building's adaptive reuse design zoning and general furniture's arrangement. Ground floor plan

Emptied of any object to ensure new functionalities, these shells will host a collection of new furniture divided into two categories. Some of them will be recovered objects from the village, others will be new contemporary articles that will fit the contemporary interior's needs . These lasts will be white elements to underline the contrast between what is original and what is not, this colour choice is also very easy to obtain in any furniture typology, and often the cheapest solution as well. Many of the contemporary furniture's additions consist in shelves or displays made of rigorously white-coated tubular metal elements, panels and joints featuring telescopic fastenings that will ensure easy assembly, low costs, maximum modularity and, more importantly, reversibility and minimum impact on the original structures.

The original materials of the rooms will be cleaned and restored when possible, white coated metal bars, plates or lime based white continuous flooring solutions should fill any gap, missing parts or demolitions, the latter will be minimized as much as possible as a general design rule of the whole masterplan.[24] Exterior façades and any window or opening will be cleaned and restored, coherently with all specifics that can be found in the preservation guidelines' chapter.

The only additions that will affect the exterior elevations will be limited to the concrete slopes introduced to ensure accessibility in every facility. A five to ten centimetres gap will be guaranteed between the old and new structure to ensure ventilation.

5.6   The House of Culture's Museum

The former House of Culture building can host a museum facility with a fair functional reuse coherence and a limited degree of demolitions and additions.

Fig. 5.21 Building 14, the House of Culture, in 2015. View from the main street

---

[24]See the Preservation Guidelines in Chap. 4 for more information.

Fig. 5.22 The House of Culture interiors' pictures shot in 2015. The main corridor (top), the former gym: abandoned for years and reactivated in 2019 (bottom)

This service will be the cultural information centre of the whole village and a variety of topics centred on the village and its whereabout can be concentrated in one single museographic experience.

These topics may include, but not be limited to: the story of Yererouyk, its excavation and its restoration interventions, the story of Anipemza, genocide information especially centred on the orphans hosted in this village, the history of the soviet town quarry whose tuff was used in several important public building throughout the

Fig. 5.23 The house of culture's theatre and auditorium stage in 2015

country. Other historical and cultural points of interest in the region may be exploited as well.

Moreover, the building may host the soviet librarian collection that is actually stored in one of the residential buildings providing proper preservation conditions and consultation opportunities (Fig. 5.21).

The former theatre is going to be preserved and reused as part of the museum and occasionally might host events organized by the locals.

Considering the modifications of the original structure there will be very few new doors' openings to ensure a convenient flow of the museum visitors while the theatre hall will be reduced as it appears too big for the present needs. The resulting area will become part of the museum, possibly hosting some medias, so a projection element has been set in the centre, using the reversible telescopic solution presented in the previous paragraph.[25] To contrast the original slope, a removable wooden structure has been set in the museum part, while, in the theatre, a wooden ramp has been inserted to ensure accessibility. A glass partition has been planned between the two generated spaces to prevent noise but at the same time allow visual permeability, so that the theatre volume itself and its original features and materials will become part of the museographic experience. A removable wooden structure will be integrated in one room as well, as the original floor and slab has been removed in the recent past. This solution will ensure full accessibility of the whole museum's itinerary and offers a fitting solution for those lighting and media projectors needed in the previously described rooms (Figs. 5.22 and 5.23).

A direct visual connection with a reception and small bookshop has been established by a partial demolition of the right wall in the entrance corridor for a better

---

[25] See Sect. 5.5 "Info-point building: a space of encounter".

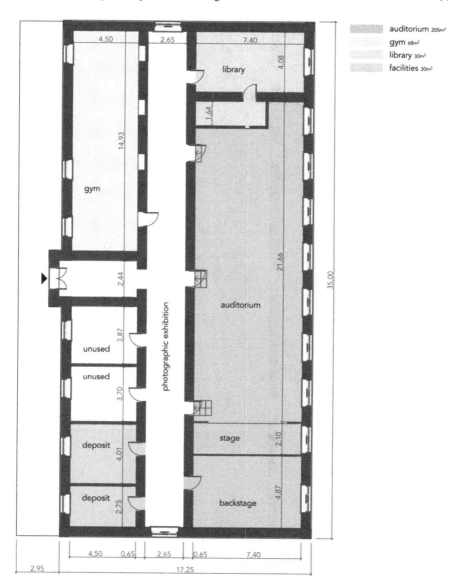

Fig. 5.24 Building 5 in 2017. Ground floor plan and existing functions' zoning

visitors' welcoming and information providing, and to take advantage of every free space in the House of Culture.

Considering the museum's outfitting, different solutions in terms of displays' configuration have been purposed. The executive solution will be opportunely

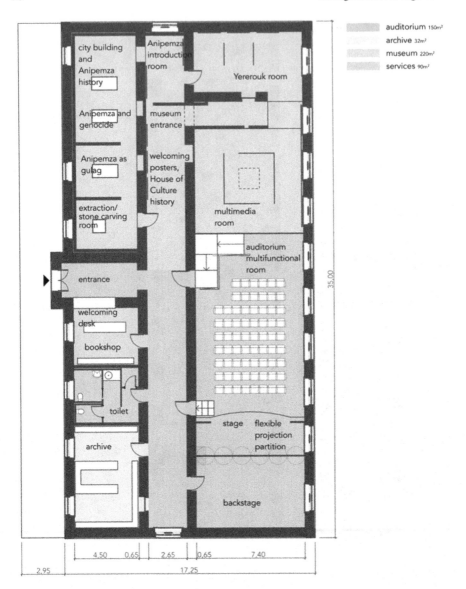

Fig. 5.25 The house of culture museum's adaptive reuse design zoning and general furniture arrangement. Ground floor plan

arranged according to the collection's characteristics. Choosing to rely on the rearrangeable telescopic tubular system described in paragraph 5.5 will provide an appreciable degree of flexibility (Figs. 5.24, 5.25 and 5.26).

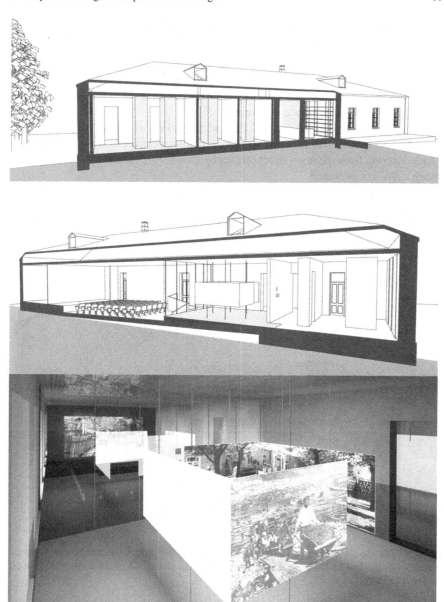

Fig. 5.26 The House of Culture's adaptive reuse design. 3D sketches and renderings. 3D section of the auditorium (top), 3D section of the Anipemza Museum (middle), rendering of the museum's multimedia area (bottom)

5.7 Temporary Sanitary Facilities' Localization

As the majority of the buildings has major plant issues or are totally lacking sanitary facilities from the original design it is therefore important to locate an area where to set a group of removable sanitary facility to ensure better conditions until all the buildings will receive adequate maintenance.[26]

An appropriate unoccupied area is located South-East of the sporting courtyard. Its central position in the village will offer a comfortable access, while the partial view's covering from the main street offered by the existing residential buildings will ensure a discreet presence.

The underground installation of Imhoff treatment basins is needed and the addition of rainwater recovery tanks might be considered as well.

Once the facility will be dismantled this area might host vegetable gardens for local food supplies (Fig. 5.27).

Fig. 5.27 The identified empty area designated for hosting temporary sanitary facilities is positioned South–West of building 10. A discreet and strategical location easily accessible from all residence buildings in the village

---

[26]See also Sect. 5.3.

5.8 The Widespread Hotel and Residential Interiors' Reorganisation

Many residential buildings' interiors have been reshaped during the course of time to fit better living standards. If the general floor plans have not been considerably altered, it is true that nowadays the majority of the apartments has become larger, including other housing units, taking advantage of the progressive abandoning of the village. To address of these necessities, at the same time trying to preserve the authenticity of the village at the highest possible level, this project suggests an example of interior remodelling that will consider a few important issues, offering a possible solution. First, the proposed design is not altering in any way the facades of the building, the position of the windows has greatly limited the remodelling for obvious natural lighting and ventilation concerns, but the integrity of the building's exteriors was one of the major concerns of our design (Figs. 5.28 and 5.29).

Second, the number of demolitions has been limited to the fewest, while additions are considered to be made using wooden frame partitions[27] that will ensure the maximum reversibility and minimum damage to the original structures. Moreover, these partitions may cover the passage of plumbing and wires, avoiding many implant upgrade issues like cuts and tracks in the walls, that would spoil forever too much of the original structure's matter. This system will also allow to work faster, lighter and cleaner, with all the consequent advantages.

Fig. 5.28 Building 18 in 2015. View from the former sporting yard and gardens. See Fig. 2.10 for a closer view of the same façade from the main street

---

[27]These wooden frame structures are actually inspired by the original soviet interior's partitions system.

Fig. 5.29 Building 18 in 2015. View from building 20, the workers' Dining Hall, the opposite façade of Fig. 5.28

Fig. 5.30 Residential building 18 in 2015. The roof's structure: detail of the staircase's volume covering

The building we considered as case study is a two-storey residential building near the Dining Hall, one of the oldest building, characterized by a considerable number of small one-room apartments with no restrooms and very large distribution corridors (Figs. 5.30, 5.31 and 5.32).

Fig. 5.31 Residential building 18 in 2015. The staircase volume from first floor's landing

Fig. 5.32 Residential building 18 in 2015. Detail of the staircase's handrail

Fig. 5.33 Residential building 18 in 2015. Ground floor corridor. The spontaneous windowed partition in this case allowed the property to take advantage of an additional window and space for their apartment, avoiding darkness in the remaining common area. With this solution original proportions of the corridor are still perceivable

Fig. 5.34 Residential building 18 in 2015. First floor corridor. Unlike the previous case reported in Fig. 5.33 the spontaneous additions completely close the corridor, significantly changing the spatial perception

Fig. 5.35 Residential building 18 in 2015. One of the common storage rooms is also used for preserves cooking and canning

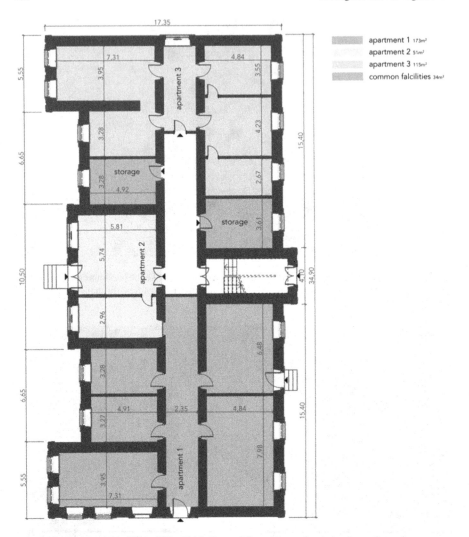

apartment 1 173m²
apartment 2 51m²
apartment 3 115m²
common falcilities 34m²

Fig. 5.36 Residential building 18 in 2015. Ground floor plan and existing layout's zoning

While trying to suggest different solutions and housing typologies to best fit the needs of the locals, the project also suggests the presence of rooms that can be rented by the locals to tourists, allowing for more revenues and possibly establishing a widespread hotel that will offer to guests different experiences in different buildings. At the same time, this solution may attract visitors promoting interaction with the locals as they will share the same buildings the inhabitants live, while respecting the privacy and comfort of the travellers.

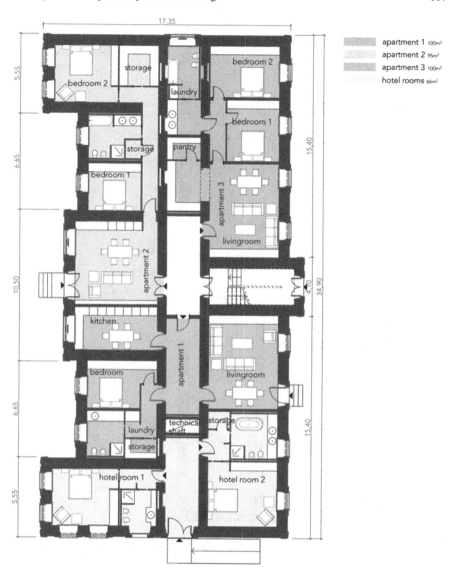

Fig. 5.37 Residential building 18 adaptive reuse design's zoning and general furniture's arrangement. ground floor plan

Since many of the original corridor's area have been added to apartments, reducing the horizontal distribution space and proportions, a glass window might be inserted in the upper part of these blocking partitions, to suggest that the original volumes of these corridors were originally longer. Residents won't be disturbed by the artificial

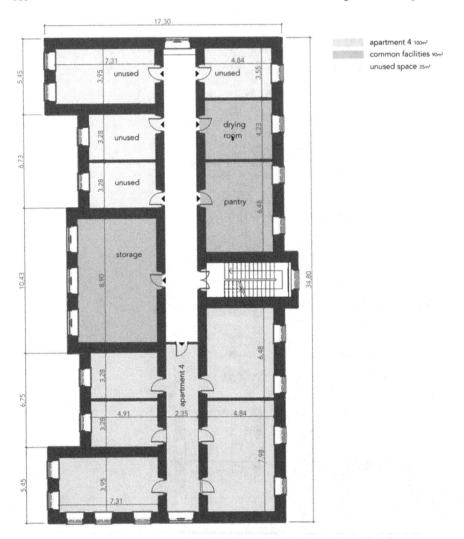

Fig. 5.38 Residential building 18 in 2015. First floor plan and existing layout's zoning

light possibly incoming from the common corridor, as these rooms are intended as entrance areas that do not communicate directly with the inner part of the apartments (Figs. 5.33, 5.34, 5.35, 5.36, 5.37, 5.38 and 5.39).

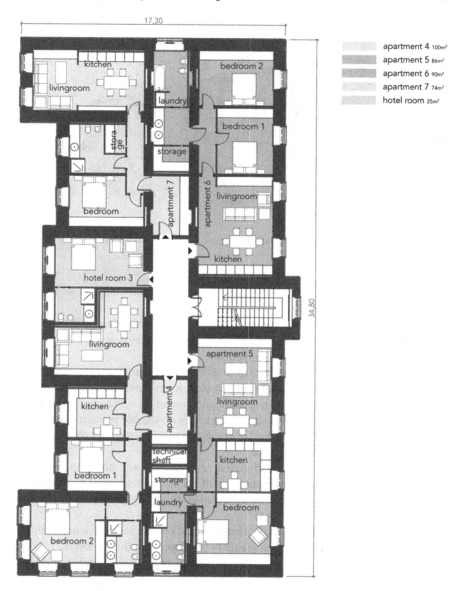

Fig. 5.39 Residential building 18 adaptive reuse design's zoning and general furniture's arrangement. First Floor Plan

5.9   A Restaurant in the Former Dining Hall

The former Soviet Dining Hall is one of the most endangered building in Anipemza. Abandon, decay and structural problems are seriously threatening its preservation.

Fig. 5.40 Building 20, the former workers' dining hall, in 2015. Northern view: main building's access

Considering the whole village, it's a unique building from many points of view and deserves to be protected.

Introducing a new restaurant appears to be a viable choice not just from the compatibility of function's point of view, but also from the general masterplan's strategies that involve the future of Anipemza.

A restaurant might offer the potential to attract more tourists as the position of the village, considering the principal tourist routes, seems compatible with a stop at lunch or dinner.[28] This may lead to a sensible increase of the revenues coming from visitors but at the same time will provide the locals a valuable place to gather and celebrate. The connection between locals' needs and visitors is again a pivotal aspect in the adaptive reuse design's proposal.

The restaurant has a strategic position. It is set at the opposite side of the village's entrance from Yererouyk. This opportunity will encourage the complete fruition of the village. A second parking[29] may be localized in the area to facilitate accessibility (Figs. 5.40, 5.41, 5.42, 5.43 and 5.44).

The interior reuse design aims to set a professional kitchen and personnel facilities in the original building as well as to create two independent dining rooms, one for larger parties, the second for more restrained groups. In these rooms all the original features will be cleaned, preserved and reintegrated only where needed, to maintain the authentic soviet times' atmosphere. All new inserted elements, like missing tables and chairs, movable partitions to enhance privacy between the tables or customize the disposition, will be white to diversify from the original context, as already mentioned in the previous reuse sections.

---

[28] See Sect. 5.2.3 "Existing major travel routes and Anipemza" for more details.

[29] A main parking area is described in Sect. 5.3.

Fig. 5.41 Building 20, the former workers' dining hall, in 2015. View from the former sporting yard

Fig. 5.42 Building 20, the former workers' dining hall, in 2015. The two main halls, featuring remains of the original soviet times' coloured plasters

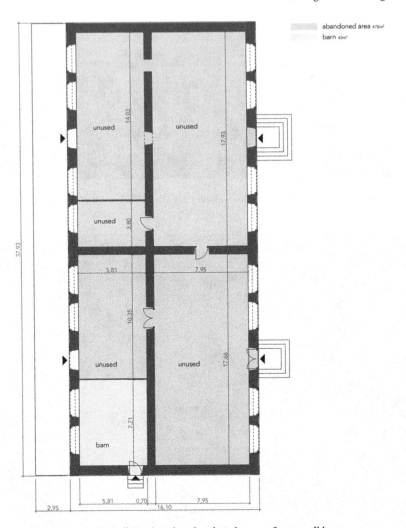

Fig. 5.43 The dining hall's building is today abandoned except for a small barn area

A set of white sound-proof plates is planned to be hanging from the ceiling to avoid refracted background noise and at the same time hide the rail-mounted spotlight sets needed to customize the lighting's solutions according to tables' disposition.

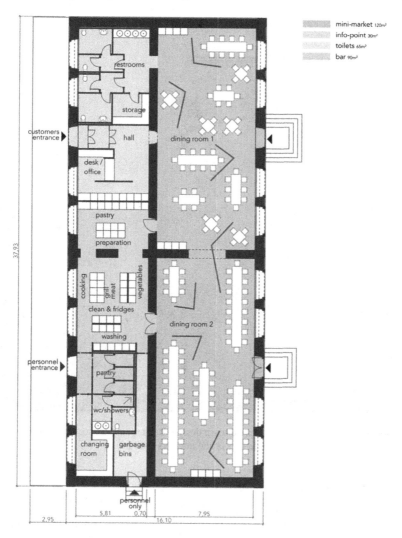

Fig. 5.44 Dining hall's adaptive reuse design's zoning and general furniture's arrangement

Since all windows of the building are missing, to underline the absence and to help HVAC system's performance, windows will be closed with fixed glasses featuring minimal frames hidden in the interior part of the windows' stone casing. White coloured wooden movable screens may be introduced to enhance the room's flow customization, while endorsing the original spatial unity of the two dining rooms (Fig. 5.45).

Fig. 5.45 Worker's dining hall's adaptive reuse design. Renderings: 3D section featuring add-on furniture highlight in grey (top), renderings of the interiors (bottom)

5.10 Conclusions

Anipemza is an interesting and unique example of a worker's village, which, over time, has been enriched by even more significant Armenian history's particularities.

Today the settlement is likely to disappear because of the tuff mining activities' recession and the consequent abandonment of the village by the residents, seeking employment elsewhere.

The abandonment is already in progress, and an even worse situation will be determined in the future if nothing will be done urgently to avoid this condition, causing the loss of this important example of Armenian urban planning and Twentieth Century's architecture.

The potentiality of Anipemza is significant and, considering that a virtuous circle is expected to be triggered, causing a turnaround in the progressive abandonment phenomena, modest investments would be necessary. Investing few economic resources for the protection of Anipemza could allow, starting by the cultural tourism's interest that Anipemza and its territory represent, to trigger a series of initiatives that would benefit the inhabitants and, more generally, Armenian tourism's offer, that certainly should still be linked to its important monuments constituted by the churches and fortresses, as well as its extraordinary landscapes, but should as well be diversified and strengthened through the testimonies of recent history, for example the preservation of the unique case of Anipemza and, in general, many other Armenian Ninetieth and Twentieth Century's architectures.

Possibly, in the future, Anipemza together with Ani (in UNESCO World Heritage tentative list since 2012) and Yererouyk (in UNESCO World Heritage tentative list since 1995) will constitute a transnational UNESCO World Heritage site, representing an important reconciliation symbol for Armenian and Turkish People.

In our opinion, this is actually a laudable dream well-worth being pursued.

References

1. Dezzi Bardeschi M (2004) Restauro: due punti e da capo. Francoangeli, Milan
2. Del Bo A, Bignami DS (eds) (2014) Sustainable social economic and environmental revitalization in Multan city. a multidisciplinary Italian–Pakistani project. Springer, Cham
3. Dezzi Bardeschi M (1991) Restauro: punto e da capo. Francoangeli, Milan, EU, Frammenti per una (impossibile) teoria
4. World Travel and Tourism Council (2015) Travel and tourism. Economic impact 2015. Armenia, London
5. Republic of Armenia Government (ed) (2014) *Armenian development strategy 2014–15*, annex to: RA Government Decree n°442-N, 27/03/2014
6. Tarlanyan H (2009) Anipemza rural community program 2013–2016 social-economic development. Anipemza Community Management Report, Anipemza
7. Berger M, Wong L (eds) (2009) Adaptive reuse, vol 1 of interventions adaptive reuse journal. Department of Interior Architecture, Rhode Island School of Design, Autumn

Suggested Readings

History of Armenia and Anipemza Village

1. Alpago Novello A (1995) The Armenians: 2000 Years of Art and Architecture, Booking International
2. Akçam T. (2005), Nazionalismo turco e genocidio armeno: dall'impero ottomano alla Repubblica, Italian edition by Antonia Arslan. Guerini, Milano
3. Armenia Illustrated Album (2009) A project of Samvel Karapetian edited by research on armenian architecture organisation. In Tigran Metz Publishing House, Yerevan
4. Augelli F, Khachatourian Saradehi A, Khachatourian Saradehi L (2015) Anipemza: from genocide orphans' village to workers village. First proposals for conservation, valorisation and improvement of an interesting architectural settlement example and of a rich history site in Armenia, Scientific Papers of National University of Architecture and Construction of Armenia IV(59):14–28. Yerevan (2015)
5. Bobelian M (2009) Children of Armenia. A forgotten genocide and the Century struggle for justice. Simon & Shuster, New York
6. Carugi U, Visone M (2017) Time frames. Conservation of policies for twentieth century architectural heritage. Routledge Taylor & Francis Group, London and New York
7. Casnati G (ed) (2014) The Politecnico di Milano in Armenia. An Italian Ministry of Foreign Affairs project for Restoration Training and Support to Local Institutions for the Preservation and Conservation of Armenian Heritage, Oemme Edizioni, Venezia
8. Cemal H (2015) 1915: genocidio armeno. Guerini e associati, Milano
9. Volli MC (ed) (2015) Il genocidio infinito: 100 anni dopo il Metz Yeghérn. Guerini e associati, Milano
10. Darieva T, Kaschuba W, Krebs M (2011) Urban spaces after socialism. Ethnographies of public spaces in Eurasian cities, Campus Verlag, Frankfurt/New York
11. Dédéyan G (ed) (2002) *Storia degli armeni*, Italian edition by Antonia Arslan and Boghos Levon Zekiyan, Guerini, Milano
12. Dipak R, Pant E, Rig-ge K (eds) (2002) Armenia, the next economy: an inter-disciplinary future-view. Crespi, Cassano Magnago
13. Dipak R (2000) The Armenian scenarios: strategic foresight of security, business and culture in the Republic of Armenia. Crespi, Cassano Magnago
14. Dundar F (2010) Crime of numbers. The role of statistics in the Armenian question (1878–1918). Transaction Publisher, New Brunswick, New Jersey
15. Flores M (2006) Il genocidio degli armeni. Il mulino, Bo-logna
16. Hovannisian R G (1971) The republic of Armenia. The first year 1918–1919. Volume I, University of California Press, London

17. Hurutynyan E (2009) The durable handwriting of Armenian. Yerevan
18. Grigoryan AG, Tovmasyan MZ (1986) Architecture of the Soviet Armenia. Mosca
19. Lewy G (2006) Il massacro degli armeni: un genocidio contro-verso. G. Einaudi, Torino
20. Marshall Lang D (1989) Armeni: un popolo in esilio. Calderini, Bologna
21. Masih JR, Krikorian RO (1999) Armenia at the crossroads. Hardwood academic publisher, Amsterdam
22. Meda F (1918) La questione armena. F.lli Treves, Milano
23. Noble J, Kohn M, Systermans D (2008) Georgia, Armenia e Azerbaigian, 3rd edn. EDT, To-rino
24. Payaslian S (2011) The political economy of human rights in Armenia. The authoritarism and democracy in a former Soviet Republic. Tauris & Co Ltd, London
25. Sirak S (2005) My memoirs. Research on Armenian Architecture Center, Yerevan
26. Tarlanyan H (2009) *Program Anipemza Rural Community 2013-2016 Social – Economic Development*, Anipemza community manager, report, Anipemza
27. Ternon Y (2003) Gli armeni: 1915–1916, il genocidio dimenticato. Rizzoli, Milano
28. Dadrian V N (2003) Storia del genocidio armeno: conflitti nazionali dai Balcani al Caucaso. Italian edition by Antonia Arslan and Boghos Levon Zekiyan, Guerini, Milano
29. Vitale S (ed) (1988) *Viaggio in Armenia*, Osip Mandel, Adelphi, Milano

Survey, Decay, Diagnosis, Structural Investigation

30. Augelli F (2006) La diagnosi delle opere e delle strutture lignee. Le ispezioni, Saonara, Il Prato
31. Blake B (2007) Metric survey techniques for historic buildings. In: Michael Forsyth (ed) Structures and construction in historic building conservation. Blackwell Publishing Ltd, Oxford
32. Building Stone Institute (2010) Recommended best practices
33. Campanella C (2017) Il rilievo degli edifici. Dario Flaccovio, Palermo
34. Duguay G (1992) The architectural preservation process. In: Heritage Notes, Architectural Preservation, Alberta Historical Resource Foundation, Alberta
35. Feilden Bernard M (2003) Conservation of Historic Buildings, pp 25–79. Elsevier, Oxford
36. Fischetti DC (2009) Structural Investigation of Historic Buildings. Wiley, New Jersey
37. Forsyth M (2008) Materials and skills for historic building conservation. Blackwell Publishing Ltd, Oxford
38. Gianbruno M, Pistidda S (ed) (2015) The walled city of Multan. Guidelines for maintenance, conservation and reuse work. Altralinea Edizioni, Firenze
39. Harris SY (2001) Building pathology: deterioration, diagnostics, and intervention. Wiley
40. Hume I (2007) Investigating, monitoring and load testing historic structures. In: Michael Forsyth (ed) Structures and construction in historic building conservation. Blackwell Publishing Ltd, Oxford
41. ICOMOS (1999) International wood committee. Principles of practice for the preservation pf historic timbers building
42. ICOMOS–ISCS (2010) Illustrated glossary on stone deteriorations patterns
43. Zevi L (ed) (2001) Il manuale del restauro architettonico. Mancosu Editore
44. ICOMOS-ISCS (2008) Illustrated glossary on stone deterioration patterns. Champigny/Marne
45. Irish Department of Culture (2004) The development plan. Record of protected structures. In: Architectural heritage protection guidelines for planning authorities, Chap. 2. Heritage Conservation, Heritage and the Gaeltacht, Dublin
46. Irish Department of Culture (2004) The development plan. Architectural conservation areas. In: Architectural heritage protection guidelines for planning authorities, Chap. 3. Heritage Conservation, Department of Culture, Heritage and the Gaeltacht, Dublin

47. Letellier R (2007) Recording, documentation, and information management for the conservation of heritage places. The Getty Conservation Institute, Los Angeles
48. May E, Jones M (2006) Conservation science. Heritage materials. The royal society of chemistry, pp 121–264. Cambridge
49. Pendlebury J, Townshend T (1998) An illustrated glossary of architectural and constructional terms: for students and newly qualified planner in practice, Section 2—building materials. School of Architecture, Planning & Landscape Global Urban Research Unit. University of Newcastle upon Tyne
50. Stylianidis E, Remondino F (2016) 3D recording, documentation and management of cultural heritage. Whittles Publishing, Scotland, UK
51. The Heritage Forum of Central Europe (2017) Heritage and the city. International Cultural Centre, Krakow
52. Travis C, McDonald J (1994) Understanding old buildings: the process of architectural investigation, 35 preservation briefs. Department of the Interior National Park Service Cultural Resources-Heritage Preservation Services, USA
53. Torraca G (2009) Lectures on materials science for architectural conservation. The Getty Conservation Institute, Los Angeles

Architectural Conservation Bibliography

54. Architectural Conservation Areas. The Development Plan (2004) Architectural heritage protection. Guidance on Part IV of the Planning and Development Act 2000, Heritage Conservation, Chap. 3. Department of the Environment, Heritage and Local Government of Ireland, Dublin
55. Assessing cultural heritage significance using the cultural heritage criteria (2013) By heritage branch. Department of Environment and Heritage Protection, Queensland
56. Augelli F (2006) La diagnosi delle opere e delle strutture lignee. Le ispezioni, Saonara (PD)
57. Walls and other structural elements (2004) Architectural heritage protection. Guidance on part IV of the planning and development act 2000, heritage conservation, Chap. 8. Department of the Environment, Heritage and Local Government of Ireland, Dublin
58. Roofs (2004) Architectural heritage protection. Guidance on Part IV of the planning and development act 2000, Heritage conservation, Chap. 9. Department of the Environment, Heritage and Local Government of Ireland, Dublin
59. Openings. Doors and Windows (2004) Architectural heritage protection. Guidance on part IV of the Planning and Development Act 2000, Heritage Conservation, Chap. 10. Department of the Environment, Heritage and Local Government of Ireland, Dublin
60. Curtilage and attendant grounds (2004) in Architectural heritage protection. Guidance on Part IV of the Planning and Development Act 2000, Heritage Conservation, Chap. 13. Department of the Environment, Heritage and Local Government of Ireland, Dublin
61. Eklund JA, Hummelt K (2013) Growing old gracefully, Information for historic building owners, Historic Scotland-Alba Aosmhor. National Conservation Centre, Longmore House, Edinburgh
62. Non-habitable protected structures (2004) in Architectural heritage protection. Guidance on Part IV of the Planning and Development Act 2000, Heritage Conservation, Chap. 14. Department of the Environment, Heritage and Local Government of Ireland, Dublin
63. Cameron S et al (1997) Biological growths on sandstone buildings. Control and treatment, printed by the stationery office on behalf of Historic Scotland and published by Historic Scotland, Edinburgh
64. Canada's Historic Places (2010) The Standards and Guidelines for the Conservation of Historic Places in Canada

65. Craw S, Hunnisett Snow J (2008) External timber doors, information for historic building owners, Historic Scotland-Alba Aosmhor. National Conservation Centre, Longmore House, Edinburgh
66. Climate change adaptation for traditional buildings (2016) Published by Historic Environment Scotland, Edinburgh
67. Cultural Heritage Guidelines. A handbook for staff and contractors (1999) Environmental Operation Unit. Walkerville
68. Curtis R (2007) Interior paint. A guide to internal paint finishes, information for historic building owners, Historic Scotland-Alba Aosmhor. National Conservation Centre, Longmore House, Edinburgh
69. Curtis R (2008) Timber floors, information for historic building owners, Historic Scotland-Alba Aosmhor. National Conservation Centre, Longmore House, Edinburgh
70. Curtis R (2008) Ventilation in traditional houses, information for historic building owners, Historic Scotland-Alba Aosmhor. National Conservation Centre, Longmore House, Edinburgh
71. Curtis R (2010) Structural joinery, Information for historic building owners, Historic Scotland-Alba Aosmhor. National Conservation Centre, Longmore House, Edinburgh
72. Curtis R (2013) Domestic chimneys and flues Information for historic building owners. Historic Scotland-Alba Aosmhor, National Conservation Centre, Longmore House, Edinburgh
73. Curtis R (2014) Ceramic tiled flooring Information for historic building owners. Historic Scotland-Alba Aosmhor, National Conservation Centre, Longmore House, Edinburgh
74. Curtis R, Jenkins M, Snow J (2014) Maintaining your home. Historic Scotland-Alba Aosmhor, Longmore House, Salisbury Place, Edinburgh
75. Curtis R, Kennedy A (2016) Damp: Causes and solutions, information for historic building owners, Historic Scotland-Alba Aosmhor. National Conservation Centre, Longmore House, Edinburgh
76. Davey A (2007) The maintenance of iron gates and railings, Information for historic building owners, Historic Scotland-Alba Aosmhor. National Conservation Centre, Longmore House, Edinburgh
77. Davey A (2007) Maintaining sash and case windows, information for historic building owners, Historic Scotland-Alba Aosmhor. National Conservation Centre, Longmore House, Edinburgh
78. External lime coatings on traditional buildings (2001) Technical advice note, n.2 technical conservation research and education division. The Scottish Lime Centre, Historic Scotland, Edinburgh
79. Gianbruno M, Pistidda S (ed) (2015) The walled city of Multan. Guidelines for maintenance, conservation and reuse work, Altralinea Edizioni, Firenze
80. Guidelines for planning authorities (2004) Architectural heritage protection. Guidance on Part IV of the Planning and Development Act 2000, Heritage Conservation. Department of the Environment, Heritage and Local Government of Ireland, Dublin
81. Guidelines on cultural heritage technical tools for heritage conservation and management, (2012) JP—EU/CoE Support to the Promotion of Cultural Diversity (PCDK), Funded by the European Union, Implemented by the Council of Europe
82. Hossam Mahdy (2017) Approaches to the conservation of Islamic cities: the case of Cairo, selected readings from ICCROM—ATHAR
83. ICCROM (2109) Guide de gestion des risques appliquée au patrimoine culturel, Canadian Conservation Institute
84. ICOMOS (1994) The Nara document on authenticity.
85. ICOMOS International Wood Committee (1999) Principles of practice for the preservation pf historic timbers building
86. ICOMOS (1987) The Washington Charter for the Conservation of Historic Towns and Urban Areas.

87. Ingval Maxwell I (2008) Structural cracks, Information for historic building owners, Historic Scotland-Alba Aosmhor. National Conservation Centre, Longmore House, Edinburgh

88. Hummelt K (2014) Micro-renewables in the historic environment. Historic Scotland-Alba Aosmhor, Longmore House, Salisbury Place, Edinburgh

89. Hunnisett Snow J (2015) Lead theft: guidance on protecting traditional buildings, Historic Scotland-Alba Aosmhor, Longmore House, Salisbury Place, Edinburgh

90. Interiors (2004) Architectural heritage protection. Guidance on Part IV of the Planning and Development Act 2000, Heritage Conservation, Chap. 11

91. Jenkins M (2001) Dry stone walls, information for historic building owners, historic Scotland-Alba Aosmhor. National Conservation Centre, Longmore House, Edinburgh

92. Jenkins M (2008) Rot in timber, information for historic building owners, Historic Scotland-Alba Aosmhor. National Conservation Centre, Longmore House, Edinburgh

93. Jenkins M, Maxwell I (2016) Roofing leadwork, information for historic building owners, Historic Scotland-Alba Aosmhor. National Conservation Centre, Longmore House, Edinburgh

94. Jessica Snow J, Torney C (2014) Lime mortars in traditional buildings, historic Scotland-Alba Aosmhor, Longmore House, Salisbury Place, Edinburgh

95. Maxwell I (2007) Cleaning sandstone. Risks and consequences, information for historic building owners, Historic Scotland-Alba Aosmhor. National Conservation Centre, Longmore House, Edinburgh

96. Maxwell I (2016) Repointing rubble stonework, Information for historic building owners, historic Scotland-Alba Aosmhor. National Conservation Centre, Longmore House, Edinburgh

97. Non-destructive investigation of standing structures (2001) Technical advice note, no 23 Technical conservation research and education division, by GBC Geotechnic Ltd. The Scottish Lime Centre, Historic Scotland, Edinburgh

98. Passed to the future: historic Scotland's policy for the sustainable management of the historic environment (2002) Historic Scotland, Edinburgh

99. Preparation and use of lime mortars (2003) Technical advice note, no 1 Technical Conservation Research and Education Division, The Scottish Lime Centre, Historic Scotland, Edinburgh

100. Report on Best Practice for Cultural Heritage Management (2012) by Culture Polis

101. Sash & case windows. A short guide for homeowners (2008) Information for traditional building owners, Published by Technical Conservation Group, Historic Scotland-Alba Aosmhor, National Conservation Centre, Longmore House, Edinburgh

102. Simpson & Brown Architects (2002) Conservation of plasterwork, Technical advice note, no. 2 Technical conservation research and education division, historic Scotland, Scottish Conservation Bureau, Edinburgh

103. Susan Macdonald S, Cheong C (2012) The role of public-private partnerships and the third sector in conserving heritage buildings, sites, and historic urban areas. The Getty Conservation Institute, Los Angeles

104. Torney C (2014) Indent repairs to sandstone ashlar masonry, information for historic building owners, Historic Scotland-Alba Aosmhor. National Conservation Centre, Longmore House, Edinburgh

105. Torney C (2014) Repointing ashlar masonry, Information for historic building owners. Historic Scotland-Alba Aosmhor, National Conservation Centre, Longmore House, Edinburgh

106. Torney C (2016) Restoration mortars for masonry repair, Information for historic building owners, Historic Scotland-Alba Aosmhor. National Conservation Centre, Longmore House, Edinburgh

107. UNESCO (2013) Operational guidelines for the implementation of the world heritage convention

108. Winter NV (2002) Design guidelines for historic districts in the city of Pasadena, California, with the Secretary of the Interior's Standards for Historic Preservation

109. Young ME et al (2003) Maintenance and repair of cleaned stone buildings (2003) Technical advice note, no 25. Technical Conservation Research and Education Division, The Scottish Lime Centre, Historic Scotland, Edinburgh

110. Yong S et al (2015) Conservation management guidelines for traditional courtyard houses and environment in the ancient city of Pingyao issued by Pingyao County Government (s.d.) Conservation Management Guidelines is initiated by Pingyao County Government and UNESCO, financed by China Cultural Heritage Foundation and Global Heritage Fund, compiled by Tongji University
111. Zevi L (ed) (2001) Il manuale del restauro architettonico, Mancosu Editore, Architectural Book and Review

Adaptive Reuse

112. Alfrey J, Putnam T (1992) The industrial heritage: managing resources and uses. Routledge, London
113. Andrieux JY (1992) Le Patrimoine Industriel. PUF, Paris
114. Baum M, Christiaanse K (eds) (2012) City as loft: adaptive reuse as a resource for sustainable urban development. GTA Publishers, Zurich
115. Bellosillo J (1988) Restoration and conversion of the monastery of Santa Maria La Real. In Domus 697, Sept 1988
116. Berger M, Wong L (eds) (2009) Adaptive reuse, vol 1 of interventions adaptive reuse journal. Department of Interior Architecture, Rhode Island School of Design, Autumn 2009
117. Bloszies C (2012) Old buildings, new designs. architectural transformations. Princeton Architectural Press, New York
118. Brooker G, Stone S (2004) Re-readings interior architecture and design principles of remodeling existing buildings. RIBA Publishing, London
119. Brooker G (2006) Infected interiors: remodelling contaminated buildings. IDEA J 2006:1–13
120. Catacuzino S (1989) Re-architecture—old buildings new uses. Thames and Hudson, London
121. Carbonara G (2011) Architettura d'oggi e restauro. Un confronto antico-nuovo, UTET, Milan
122. Carter DK (ed) (2016) remaking post-industrial cities: lessons from North America and Europe. Routledge, New York
123. Choffel-Mailfert MJ, Romano J (1991) Vers une transition culturelle? sciences et techniques en diffusion: patrimoines reconnus, Cultures Menacées. PUN, Nancy
124. Cowie J, Heathcott J (2003) Beyond the ruins: the meanings of deindustrialization. Cornell University Press, New York
125. Cramer J, Breitling S (2007) architecture in existing fabric. Birkhäuser, Berlin
126. Cunnington P (1988) Change of use: the conversion of old buildings. Alpha Books, London
127. Dambron P (2004) Patrimoine industriel et développement local: le patrimoine industriel et sa réappropriation territoriale. Delaville, Paris
128. Del Bo A, Bignami DS (eds) (2014) Sustainable social economic and environmental revitalization in Multan City. A multidisciplinary Italian-Pakistani Project. Springer, Cham
129. Dezzi Bardeschi M (1991) Restauro: punto e da capo. Frammenti per una (impossibile) teoria. Francoangeli, Milan, EU
130. Dezzi Bardeschi M (2004) Oltre la conservazione: il progetto del nuovo per il costruito. Ananke 42:82–85
131. Dezzi Bardeschi M (2004) Restauro: due punti e da capo. Francoangeli, Milan
132. Diamond R, Wang W (1995) On continuity. Chronicle Books, San Francisco
133. Douglas J (2006) Building Adaptation. Elsevier, Oxford
134. Fawcett W (2012) Flexible strategies for long-term sustainability under uncertainty. In: Building research and information, vol 40
135. Folke C, Carpenter S, Elmqvist T, Gunderson L, Holling CS, Walker B (2002) Resilience and sustainable development: building adaptive capacity in a world of transformations. AMBIO: J Human Environ 32:437–440

136. Giebeler G, Fisch R, Krause H, Musso F, Petzinka K, Rudolphi A (2009) Refurbishment manual: maintenance, conversions. Extensions, Birkhäuser, Basel
137. Gorgolewski M (2018) Resource salvation. The architecture of reuse. Ryerson University Toronto, Wiley Blackwell, Oxford
138. Gould R, Schiffer (eds) Modern material culture: the archaeology of us. Academic Press, New York
139. Grimmer AE, Hensley JE, Petrella L, Tepper AT (2011) Illustrated guidelines on sustainability for rehabilitating historic buildings. The Secretary of the Interior's Standards for Rehabilitation, U.S. Department of the Interior, National Park Service, Technical Preservation Services, Washington, D.C.
140. Henehan D, Woodson D, Culbert S (2004) Building change of use: renovating, adapting and altering commercial institutional and industrial properties. McGraw-Hill, New York
141. Office Heritage (ed) (2008) New uses for heritage places. NSW Department of Planning and the Royal Australian Institute of Architects, Parramatta
142. Highfield D (1990) The rehabilitation and re-use of old buildings. Spon Press, Abingdon
143. Hong Kong Building Department (2012) Practice guidebook on compliance with building safety and health requirements under the buildings ordinance for adaptive re-use of and alteration and addition works to heritage buildings. Hong Kong Buildings Department release, Hong Kong
144. Houben F (2001) Composition, contrast, complexity. Birkhauser, Basel
145. Hudson K (1975) Exploring our industrial past. Hodder & Stoughton, London
146. ICOMOS (1994) The Nara document on authenticity in relation to the world heritage convention. ICOMOS, Nara
147. Jäger F (ed) (2010) Old and new: design manual for revitalizing existing buildings. Birkhäuser, Basel
148. James P (2015) Urban sustainability in theory and practice: circles of sustainability. Routledge, London
149. Machado R (1976) Old buildings as palimpsest—toward a theory of remodelling. In: Progressive architecture. Nov 1976
150. Mintzberg H (1994) The Rise and Fall of Strategic Planning. Prentice Hall Europe, Glasgow
151. Mostafavi M, Leatherbarrow D (1993) On weathering: the life of buildings in time. MIT Press, Cambridge
152. National Trust for Historic Preservation (1980) Old and new architecture: design relationships. The Preservation Press, Washington DC
153. Orange H (2016) Reanimating industrial spaces: conducting memory work in post-industrial societies. Routledge, New York
154. Pellegrini PC (2018) Manuale del Riuso Architettonico. Dario Flaccovio Editore, Palermo
155. Perez De Arce R (1978) Urban transformations and the architecture of additions. In: Architectural design, no 4
156. Plevoets B, Van Cleempoel K (2011) Adaptive reuse as a strategy towards conservation of cultural heritage: a literature review. In: Brebbia CA, Binda L (eds) Structural studies, repairs and maintenance of heritage architecture. WIT Press, Southampton
157. Powell K (1999) Architecture reborn: converting old buildings for new uses. Rizzoli International, New York
158. Robert P (1989) Adaptations: new uses for old buildings. Princeton Architectural Press, New York
159. Robiglio M (2019) Re-USA. 20 American stories of adaptive reuse. Jovis, Berlin
160. Scott F (2008) On altering architecture. Routledge, London
161. Schittich C (ed) (2003) Building in existing fabric: refurbishment, extensions. New Design, Birkhäuser, Basel
162. Shipley R, Utz S, Parsons M (2006) Does adaptive reuse pay? A study of the business of building renovation in Ontario, Canada. Int J Herit Stud 12(6):505–520
163. Söderbaum P (1996) Economics and ecological sustainability. An actor-network approach to evaluation. Third International Workshop on Evaluation in Theory and Practice, London

164. The Royal Australian Institute of Architects (2004) Adaptive reuse. Preserving our past, building our future. Australian Government, Department of the Environment and Heritage, Canberra
165. UNESCO (2008) Best practices on social sustainability in historic districts. UNHABIT, Nairobi
166. Contributors Various (1980) Old and new architecture: design relationship. The Preservation Press, Lafayette
167. Wong L (2016) Adaptive reuse: extending the lives of buildings. Birkhäuser, Basel

History of Armenia

168. https://www.nationsonline.org/oneworld/map/armenia_map.htm
169. http://america.aljazeera.com/multime-dia/2015/4/for-last-armenian-village-in-turkey-no-emembrance-of-things-past.html
170. http://kuow.org/post/last-armenian-village-tur-key-keeps-silent-about-1915-slaughter
171. http://en.wataninet.com/features/interviews/what-if-turkey-recognises-the-genocide/13608/
172. https://it.pinterest.com/arzanita/the-armenian/
173. http://www.tempi.it/il-lungo-silenzio-degli-inno-centi-antonia-arslan-racconta-la-tormen tata-sco-perta-della-sua-eredita-armena#.VUOkICHtmkp
174. http://www.telegraph.co.uk/news/worldnews/europe/turkey/11373115/Amal-Clooneys-la-test-case-Why-Turkey-wont-talk-about-the-Ar-menian-genocide.html
175. http://mediachecker.wordpress.com/2013/06/07/the-armenian-genocide/
176. http://www.eastroute.com/armenia/museu-ms-of-armenia/museum-alexander-tamanyan
177. http://gulaghistory.org/nps/
178. http://whc.unesco.org/en/tentativelists/10/
179. https://whc.unesco.org/en/list/1518
180. http://net.lib.byu.edu/%7Erdh7/wwi/1915/bryce/
181. http://shirak.gov.am/about-communities/588/
182. http://shirak-agro.am/index.php?id=3947#.U7kFTvmSz4b
183. http://www.caa.am/arm/lg.php?section=COMMUNITIES&id=663
184. http://www.yotverk.am/datei/070231/s070202.htm
185. http://eurasia.travel/armenia/cities/northern_armenia/jrapi__anipemza/
186. http://www.caa.am/arm/lg.php?section=COMMUNITIES&id=663
187. http://www.azatutyun.am/media/video/25139641.html
188. http://gulaghistory.org/nps/onlineexhibit/stalin/perm36.php

Conservation

189. http://www.meda-corpus.net/eng/gates/visit/atl_frn.htm
190. http://www.meda-corpus.net/eng/gates/visit/ats_eng.htm
191. http://www.nps.gov/history/hps/tps/standguide/index.htm
192. http://www.iccrom.org

Anipemza Community

193. http://shirak.gov.am/about-communities/588/
194. http://shirak-agro.am/index.php?id=3947#.U7kFTvmSz4b
195. http://www.caa.am/arm/lg.php?section=COMMUNITIES&id=663
196. http://www.yotverk.am/datei/070231/s070202.htm
197. http://eurasia.travel/armenia/cities/northern_armenia/jrapi__anipemza/
198. http://shirak.gov.am/about-communities/588/
199. http://www.caa.am/arm/lg.php?section=COMMUNITIES&id=663
200. http://www.azatutyun.am/media/video/25139641.html

Tour Operators

201. http://www.tourismarmenia.org
202. http://adventure.nationalgeographic.com
203. http://www.welcomearmenia.com
204. http://www.armeniapedia.org
205. http://europetravelz.com/
206. http://www.visitarmenia.org/
207. http://www.advantour.com/armenia/
208. http://www.fructusarmeniacus.com/
209. http://wikitravel.org/en/Armenia
210. http://www.armeniainfo.am/
211. http://www.deagostinigeografia.it/wing/schedapaese.jsp?idpaese=010
212. http://wikitravel.org/en/Armenia_in_9_Days
213. http://www.armeniapedia.org
214. http://www.tourismarmenia.org
215. http://www.cilicia.com/armo5_clickmap.html
216. http://www.travel.am
217. http://www.araratour.com
218. http://www.visavistour.com
219. http://anitour.am/it/
220. https://www.biketours.com
221. http://www.atb.am/it/
222. http://www.ecotourismarmenia.com

Sociology and Tourism's Marketing

223. http://www.armstat.am
224. http://mediaau.com/starter-personas-travel-content/
225. http://www.wttc.org
226. http://contentmarketinginstitute.com
227. http://www.euromonitor.com

Index

© The Author(s), under exclusive license to Springer Nature Switzerland AG 2021
F. Augelli et al., *Preservation and Reuse Design for Fragile Territories'*
Settlements, SpringerBriefs in Applied Sciences and Technology,
https://doi.org/10.1007/978-3-030-45497-5

Printed in the United States
By Bookmasters